The Best American Science Writing 2005

Merry Christmas,

John —

may these essays
ignite your mind and
provide new insights
& wisdom, Love
Holley Donovan

THE BEST AMERICAN SCIENCE WRITING

The Best American
SCIENCE WRITING
2005

EDITOR: ALAN LIGHTMAN

Series Editor: Jesse Cohen

HARPER **PERENNIAL**

HARPER ● PERENNIAL

Permissions appear following page 300.

THE BEST AMERICAN SCIENCE WRITING 2005. Compilation copyright © 2005 by HarperCollins Publishers. Introduction copyright © 2005 by Alan Lightman. All rights reserved. Printed in the United States of America. No part of this book may be used or reproduced in any manner whatsoever without written permission except in the case of brief quotations embodied in critical articles and reviews. For information address HarperCollins Publishers, 10 East 53rd Street, New York, NY 10022.

HarperCollins books maybe purchased for educational, business, or sales promotional use. For information please write: Special Markets Department, HarperCollins Publishers, 10 East 53rd Street, New York, NY 10022.

FIRST EDITION

Designed by Cassandra J. Pappas

Library of Congress Cataloging-in-Publication Data is available upon request.

ISBN-10: 0-06-072642-3

ISBN-13: 978-0-06-072642-3

05 06 07 08 09 BVG/RRD 10 9 8 7 6 5 4 3 2 1

Contents

Introduction by Alan Lightman

WHEN I WAS SIXTEEN, a frightened aunt gave me a copy of a new book called *The Scientific Endeavor*. The aunt was not unfamiliar with my wayward scientific investigations—on one occasion, I had caused a small fire in the house with an unfortunate mix of chemicals; on another, I had knocked out the television reception in my neighborhood with a Tesla coil run off a six-volt battery—and I suspect she was hoping that I would expend my raging science drive on reading rather than on experimenting. Her wishes were granted. I was completely mesmerized by this book. Published on the occasion of the centennial celebration of the U.S. National Academy of Sciences, in 1963, *The Scientific Endeavor* contained addresses by two dozen Nobel Prize–caliber scientists holding forth on their disciplines. And the chapters could be partly digested by a high school student. Here was the biologist George Wald writing about "The Origin of Life," chemist Linus Pauling writing about "The Architecture of Molecules," and physicist William Fowler writing about "The Origin of the Elements." (Unbeknowst to me at the time, six years later I would take a bugle-call, eight-a.m. class in nuclear astrophysics from Fowler at Caltech, a class of such joyous intensity that even when an earthquake severely shook the building one morning, Fowler continued chalking equations on the blackboard and we graduate students remained glued to our seats.) This writing thrilled me. With the help of these expert guides, I was taken to the frontier, standing on the precipice of the known and the unknown, and I felt a rush of blood through my body.

The articles in *The Scientific Endeavor* were well written, but they were not quite aimed toward the general public. In fact, the golden age of public science writing in America had not yet arrived. (I prefer the term "public science writing" to "popular science writing," as the latter has too much a whiff of the commercial or the diluted-to-appeal-to-the-masses.) The first glimmerings of that golden age could already be seen in the beautiful essays of Loren Eiseley, whose *Darwin's Century* was published in 1958, and in the lyrical yet powerful writing of Rachel Carson, whose *Silent Spring* (1962) was first serialized in *The New Yorker*. However, these great science pieces were few and far between. Moreover, there was at this time a strong prejudice within the scientific community that public science writing was an activity much inferior to the actual doing of science and that scientists should not waste their precious time on "popularizing" their subject for the public.

Over the next fifteen years, the landscape began to tilt. In 1968, James Watson published *The Double Helix*. Carl Sagan's *The Cosmic Connection* arrived in 1973. When I went to Cornell in 1974, as a postdoctoral fellow in astrophysics, I remember being very much aware that Sagan was there, writing about science for the public. I was also aware that, for these public indelicacies, he was held in some disrepute, like an otherwise distinguished uncle who is discovered to have a trace of "bad blood" on his side of the family. I wanted to do what the ebullient Sagan did, although our styles were vastly different. Also around this time, Lewis Thomas began writing his highly accessible essays in the *New England Journal of Medicine*, culminating in his National Book Award winning collection *The Lives of a Cell* in 1974. Slightly later, Stephen Jay Gould initiated his long-running series of essays for *Natural History* magazine, with his first collection *Ever*

Since Darwin appearing in 1977. Nobel Prize–winning physicist Steven Weinberg's lovely book *The First Three Minutes* arrived to the delight of the public also in 1977, followed by Freeman Dyson's *Disturbing the Universe* in 1979. A few years later, David Quammen started his witty columns on natural history for *Outside* magazine. Meanwhile, in 1978, the *New York Times* inaugurated its splendid weekly section on science, "Science Times." The times were surely changing.

By the time I began writing a monthly essay on science for *Science 80* in 1982, public science writing by scientists was just beginning to be a legitimate and valued activity. The late 1980s and 1990s, of course, have witnessed an explosion of public science writing, by scientists and nonscientists alike. Now, the golden age is here. The poet Rainer Maria Rilke wrote that we should "try to love the questions themselves, like locked doors and like books written in a very foreign tongue," but good public science writing has enabled us to love the answers as well as the questions, and to receive those answers in language that we can understand.

There are as many kinds of science writing as there are of writing in general, each with its own tools and terrain. There is reportage, in which the writer "gets the story," interviews the experts, and largely stays out of the way. There is the essay, in which the writer goes inward rather than outward, casting him or herself center stage and unashamedly inviting the reader to watch as the writer personally grapples with an idea. Yet a third category might be called experimental narrative. Here, the writer may be trying to capture a scene or a moment of life, as in fiction writing, without full explanation or understanding, or may be constructing a fantasy that demonstrates important principles of science. Examples of this genre include Primo Levi's incomparable *The Periodic Table*, Edwin Abbott's classic *Flatland*, or George Gamow's Mr. Tompkin's books.

Like all writing, the best science writing is clear, captivating, intelligent, provocative, imaginative, graceful, and funny when the humor is natural. (Oh, that my own writing could always live up to these ideals.) There are no rules. As Henry James said in "The Art of Fiction," the only rule that cannot be broken is that the writing must be interesting. But I believe that science writing has some additional demands over and above other forms of writing, first because the subject matter is often technical, and second because science terrifies some people.

To deal with the first challenge, I do not advise skirting the difficult parts of the subject. Rather, get your hands dirty. Talk to experts. Study. Learn the material well enough so that you can make simplifications while maintaining the essential ideas. Here, good metaphors are of high premium. For example, how can a meager layperson possibly conceive of the fundamental concept in cosmology that the universe is expanding, with all the galaxies moving away from each other, and yet with no center to the expansion? In 1931 the astronomer Arthur Eddington said to imagine that space is two-dimensional and that the galaxies are dots painted on the surface of an expanding balloon. From the vantage of any one dot, the other dots are moving away from it in all directions, yet no dot is the center. This powerful metaphor has helped students of cosmology ever since, in every language where the subject is explained.

A prime strategy for dealing with the second challenge, the "fear of science," is to portray the human side of science. No matter how seemingly remote its subject matter, the enterprise of science is a profoundly human endeavor. "Only connect!" wrote E. M. Forster in *Howards End*. Scientists have passions, fears, jealousies, guilt, ambitions, and moments of nobility and cowardice like everyone else. Scientists have life stories and human dramas.

Consider string theory, possibly the most arcane area of science today. String theory concerns ultra-small vibrating bits of matter, far smaller than the nucleus of an atom. (A few extra dimensions of space besides the usual three are aslo involved, but that's a detail.) Twenty years ago, when theoretical physicist John Schwarz first proposed string theory, he jumped on a stage at the Aspen Center for Physics and, in the words of science writer Dennis Overbye, "began babbling about having discovered a theory that could explain everything." According to a prearranged plan, men in white suits then appeared from the wings and carried Schwarz away. A few years later, when the laughing had stopped, Schwarz realized that his string theory could indeed explain all the forces of nature. "I was immediately convinced this was worth devoting my life to," Schwarz recalled in his interview with Overbye. So far, not a shred of experimental evidence supports string theory. However, some of the best theoretical physicists in the world are infatuated with it. That's a story. Roger Straus of Farrar, Straus, and Giroux once told me that the secret of great science writing, or any other kind of writing, is to tell a story. "People always want to hear a story," said Roger.

The articles and essays I have chosen here illustrate, in one way or another, all the ideals of fine science writing. I would never claim that these pieces are *the best* science writing published in 2004. But they are all excellent examples of the craft and bring pleasure to me when I read them. It is my hope that these pieces, and others like them, will excite the imaginations of sixteen-year-olds, just as another book did for me long ago.

OLIVER SACKS

Greetings from the Island of Stability

FROM THE *NEW YORK TIMES*

As the neurologist and writer Oliver Sacks revealed in his charming memoir Uncle Tungsten, *his first love was chemistry, especially the elements that make up the periodic table. The announcement of the discovery of two new elements allows Sacks to revive his childlike enthusiasm and astonishment for the wonders of chemistry.*

On February 2, 2004, the discovery of two new elements—113 and 115—was announced by a team of Russian and American scientists. There is something about such announcements that raises the spirits, thrills one, evokes thoughts of new lands being sighted, of new areas of nature revealed.

It was only at the end of the eighteenth century that the modern idea of an "element" was clearly defined, as a substance that could not be decomposed by any chemical means. In the first decades of the nineteenth century, Humphry Davy, the chemical equivalent of a big-game hunter, thrilled scientists and public alike by bagging potassium, sodium, calcium, strontium, barium, and a few other elements. Discoveries rolled on throughout the next one hundred years, often exciting the public imagination, and when, in the 1890s, five new elements were discovered in the atmosphere, these quickly found their way into H. G. Wells's novels—argon was used by the Martians in *The War of the Worlds*; and helium to make the antigravity material that transported Wells's heroes in *The First Men in the Moon.*

The last naturally occurring element, rhenium, was discovered in 1925. But then, in 1937, there came something no less thrilling: the announcement that a new element had been *created*—an element that seemingly did not exist in nature. The element was named "technetium," to emphasize that it was a product of human technology.

It had been thought that there were just ninety-two elements, ending with uranium, whose massive atomic nucleus contained no less than ninety-two protons, along with a considerably larger number of neutral particles (neutrons). But why should this be the end of the line? Could one create elements beyond uranium, even if they did not exist in nature? When Glenn T. Seaborg and his colleagues at the Lawrence Berkeley National Laboratory in California were able to make a new element in 1940 with ninety-four protons in its huge nucleus, they could not at first imagine that anything more massive would ever be obtained, and so they called their new element "ultimium" (later it would be renamed plutonium).

If such elements with enormous atomic nuclei did not exist in nature, this was, presumably, because they were too unstable: with more and more protons in the nucleus repelling each other, the nucleus would tend toward spontaneous fission. Indeed, as Seaborg and his colleagues strove to make heavier and heavier elements (they created nine new ones over the next twenty years, and Element 106 is now named seaborgium in his honor), they found that these were increasingly unstable, some of them breaking up within microseconds of being made. There seemed good grounds for supposing that one might never get beyond Element 108—that this would be the absolute "ultimium."

Then, in the late 1960s, a radical new concept of the nucleus emerged—the notion that its protons and neutrons were arranged in "shells" (like the "shells" of electrons that whirled around the nucleus). The stability of the nucleus of an atom, it was theorized, depended on whether these nuclear shells were filled, just as the chemical stability of atoms depended on the filling of their electron shells. It was calculated that the ideal (or "magic") number of protons required to fill such a nuclear shell would be 114, and the ideal number of neutrons would be 184. A nucleus with both these numbers, a "doubly magic" nucleus, might be, despite its enormous size, remarkably stable.

This idea was startling, paradoxical—as strange and exciting as that of black holes or dark energy. It moved even sober scientists like Seaborg to allegorical language. He thus spoke of a sea of instability—the increasingly and sometimes fantastically unstable elements from 101 to 111—that one would somehow have to leap over if one was ever to reach what he called the island of stability (an elongated island stretching from Elements 112 to 118, but having in its center the "doubly magic" isotope of 114). The term "magic" was continually used—Seaborg and others spoke of a magic ridge, a magic mountain and a magic island of elements.

This vision came to haunt the imagination of physicists the world over. Whether or not it was scientifically important, it became psychologically imperative to reach, or at least to sight, this magic territory. There were undertones of other allegories as well—the island of stability could be seen as a topsy-turvy, Alice-in-Wonderland realm where bizarre and gigantic atoms lived their strange lives. Or, more wistfully, the island of stability could be imagined as a sort of Ithaca, where the atomic wanderer, after decades of struggle in the sea of instability, might reach a final haven.

No effort or expense was spared in this enterprise. The vast atom-smashers, the particle colliders of Berkeley, Dubna, and Darmstadt were all enlisted in the quest, and scores of brilliant workers devoted their lives to it. Finally, in 1998, after more than thirty years, the work paid off. Scientists reached the outlying shores of the magic island: they were able to create an isotope of 114, albeit nine neutrons short of the magic number. (When I met Glenn Seaborg in December 1997, he said that one of his longest-lasting and most cherished dreams was to see one of these magic elements—but, sadly, when the creation of 114 was announced in 1999, Seaborg had been disabled by a stroke, and may never have known that his dream had been realized.)

Since elements in the vertical groups of the periodic table are analogous to one another, one can say with confidence that one of the new elements, 113, is a heavier analog of Element 81, thallium. Thallium, a heavy, soft, leadlike metal, is one of the most peculiar of elements, with chemical properties so wild and contradictory that early chemists did not know where to place it in the periodic table. Indeed, it was sometimes called the platypus of elements. Is thallium's new, heavier analog, "super-thallium," as strange?

Similarly, the other new element, 115, is certain to be a heavier analog of Element 83, bismuth. As I write, I have a lump of bismuth in front of me, prismatic and terraced like a miniature Hopi village, and glittering with iridescent oxidation colors, and I cannot help wondering whether "super-bismuth," if it could be obtained in massive form, would be as beautiful—or perhaps more so.

And it could be possible to obtain more than a few atoms of these super-heavy elements, for they may have half-lives of many years; unlike the elements preceding them, which vanish in split seconds. For instance, atoms of Element 111, the heavier analog of gold, break down in less than a millisecond, and it is difficult to have more than an atom or two at a time, so we may never hope to see what "super-gold" looks like. But if we can make isotopes of 113, 114 (super-lead), and 115, which may have half-lives of years or centuries, we will have three enormously dense and strange new metals.

Of course, we can only guess at what properties 113 and 115 will possess. One can never tell in advance what the practical use or scientific implications of anything new might be. Who would have thought that germanium—an obscure "semi-metal" discovered in the 1880s—would turn out to be crucial to the development of transistors? Or that elements like neodymium and samarium, regarded for a century as mere curiosities, would turn out to be essential to the making of unprecedentedly powerful permanent magnets?

But such questions are, in a sense, beside the point. We search for the island of stability because, like Mount Everest, it is there. But, as with Everest, there is profound emotion, too, infusing the scientific search to test a hypothesis. The quest for the magic island shows us that science is far from being coldness and calculation, as many people imagine, but is shot through with passion, longing, and romance.

JAMES GLEICK

Isaac Newton's Gravity

FROM *SLATE*

An exhibition of Newton's life and work is the occasion for James Gleick to reflect on the man behind the myth.

A curious thing happened to Isaac Newton on the way to a grand new exhibition at the New York Public Library, The Newtonian Moment: Science and the Making of Modern Culture. He seems to have gone through a time machine—backward.

On display is what must be the most impressive collection of Newtoniana ever assembled in the United States: early notebooks and manuscript fragments, antique scientific instruments and rare books, portraits and at least one death mask. Even looking through quarter-inch Plexiglas, you can feel the power of the books and papers. Especially awesome are the rare editions of the *Philosophiae Naturalis Principia Mathematica,* the book that revealed Newton's System of the World, and the notebook pages covered edge to edge with the words and figures that flowed from Newton's quill in his astonishingly tiny and elegant hand.

But it's a nineteenth-century Newton who's been dusted off here: Sir Isaac, powdered and bewigged, the genius of rationality and order, who created—and who came to personify—modern science. "The acme of human possibility," in the words of the exhibit's curator, Mordechai Feingold, a history professor at the California Institute of Technology. Up to a point, this is fine. Newton really did write down the rules of the universe we live in, a universe of science and industry and reason, in which humans have managed to achieve a fair degree of control over unruly Nature.

Still, it is only a partial, bowdlerized picture of the man. We know much more now, thanks to long-buried papers that began coming to light during the twentieth century. We know about Newton's pathological aloneness, his brush with madness, his obsession with alchemy and theological heresies—none of which is so much as hinted at in this exhibition, let alone explored.

When Newton died in 1727, at the age of eighty-four, he was already an iconic figure, celebrated in verse and portraiture. His ornate tomb at Westminster Abbey was inscribed, "Mortals rejoice that there has existed so great an ornament of the human race." But most of the millions of words he had written in his lifetime remained hidden. His passionate religious convictions were a dangerous secret: With all his heart he disbelieved in the Holy Trinity, and this was heresy. He enjoyed the fame and riches that came to him, but for most of his life he had preferred seclusion to participation in England's burgeoning scientific community.

Newton never married, apparently never had a lover, and never even had a real friend, as we use the word in our sociable times. He never had a scientific collaborator; indeed, he fought bitterly and ruthlessly with other great philosophers. Having been a fellow and professor at Trinity College, Cambridge, for most of his adult life, he left behind not a single person who claimed to have been his student. When he made his greatest discoveries, his instinct was not to publish them but to keep them to himself.

Newton's legacy is more than the sum of his discoveries. His flaws, his errors, and his scheming, too, changed the direction of science in profound ways. A case in point: Visitors streaming through the

library's Gottesman Exhibition Hall may stop to examine two peculiar pages, faded and stained, regarding a dispute over who invented
the calculus. But they aren't likely to realize that these pages are the
damning evidence—the smoking gun—of one of the most delicious
frauds in the annals of science.

Newton had, in fact, invented most of the mathematical machinery we now call the calculus. He accomplished this in the 1660s, as a
very young man alone in a farmhouse during the plague years, and
revealed it to no one. Meanwhile, in Germany, Gottfried Wilhelm
Leibniz also invented the calculus—which is to say, much of the same
mathematics, though with a different emphasis and a different form
of notation. Leibniz's form is the one we use today. Leibniz was
entirely happy to publicize his discovery, and by 1712, when they were
both old men, he and Newton were embroiled in an ugly international dispute, each accusing the other of outright theft.

The Royal Society of London, with Sir Isaac presiding, appointed
a committee of scholars to adjudicate the matter. Their report
found no doubt whatsoever. It vindicated Newton and con-
demned Leibniz. In addition to the report itself, the Royal Society
published an anonymous review of the report, and this, too, right-
eously denounced Leibniz. "It lies upon him, in point of Candor," it
declared, "to make us understand that he pretended to this Antiquity
of his Invention with some other Design than to rival and supplant
Mr. Newton."

Candor indeed! We now understand, from the surviving hand-
written drafts, that the author of the report and the author of the
review were the same man—Newton—writing about himself in the
third person.

Before this sorry business, Leibniz himself had said, "Taking mathematics from the beginning of the world to the time when Newton
lived, what he had done was much the better half." Nothing learned
by modern science diminishes Newton's glory—not relativity or
quantum mechanics or chaos theory. His fingerprints mark every
part of science. He does not need debunkers.

So, the library offers a portrait on the model of David Brewster's 1831 hagiography: "Neither the partiality of rival nations, nor the vanity of a presumptuous age, has ventured to dispute the ascendancy of his genius," Brewster wrote. These words are still true today.

But we misunderstand Newton if we imagine him as a paragon of rationality and public science in the modern style. Brewster also wrote, "There is no reason to suppose that Sir Isaac Newton was a believer in the doctrines of alchemy," and these words don't hold up so well. Not only was he a believer; he was, in secret, the most complete and knowledgeable alchemist of his time.

If there was one turning point in modern Newton scholarship, it came in 1936, when a metal trunk of Newton's manuscripts arrived at Sotheby's in London. These notebooks and loose pages, amounting to three million unread words, were to be sold at auction in hundreds of individual lots. John Maynard Keynes, horrified by this sacrilege, was able to buy some lots immediately and gather others later; the rest were scattered around the world.

What Keynes found in these manuscripts amazed him: ethereal spirits, a secret fire pervading matter, a fixation on quicksilver— mercury—as "the masculine and feminine semens . . . fixed and volatile, the Serpents around the Caduceus, the Dragons of Flammel." We know now that Newton, the alchemist, hid behind a pseudonym, *Jeova sanctus unus,* as he slowly and unwittingly poisoned himself with the mercury he continually touched, smelled, and tasted.

None of this work led anywhere, as far as modern science is concerned, and none of it is reflected in the exhibit. Still, it's a part of the real Isaac Newton, a complex and tormented soul from a pre-Newtonian world. The birth of science was messier than you would think from *The Newtonian Moment,* and more interesting, too.

This is why Keynes, speaking not long before his own death, tried to persuade us not to think of Newton as "the Sage and Monarch of the Age of Reason" on display at the library; rather, as an "intense and flaming spirit."

"Newton was not the first of the age of reason," Keynes said.

"He was the last of the magicians, the last of the Babylonians and Sumerians, the last great mind which looked out on the visible and intellectual world with the same eyes as those who began to build our intellectual inheritance rather less than ten thousand years ago." Newton opened a door to our world, sure. But he belonged to the world we have left behind.

FRANK WILCZEK

Whence the Force of F = ma?

FROM *PHYSICS TODAY*

It may seem strange for a Nobel Prize–winning quantum theorist to admit that there is a part of fundamental physics he once had trouble understanding. But Frank Wilczek uses his perplexity to discuss the deeper significance of Newton's second law of motion.

When I was a student, the subject that gave me the most trouble was classical mechanics. That always struck me as peculiar, because I had no trouble learning more advanced subjects, which were supposed to be harder. Now I think I've figured it out. It was a case of culture shock. Coming from mathematics, I was expecting an algorithm. Instead I encountered something quite different—a sort of culture, in fact. Let me explain.

Newton's second law of motion, $F = ma$, is the soul of classical mechanics. Like other souls, it is insubstantial. The right-hand side is the product of two terms with profound meanings. Acceleration is a purely kinematical concept, defined in terms of space and time. Mass quite directly reflects basic measurable properties of bodies (weights, recoil velocities). The left-hand side, on the other hand, has no independent meaning. Yet clearly Newton's second law is full of meaning, by the highest standard: It proves itself useful in demanding situations. Splendid, unlikely looking bridges, like the Erasmus Bridge (known as the Swan of Rotterdam), do bear their loads; spacecraft do reach Saturn.

The paradox deepens when we consider force from the perspective of modern physics. In fact, the concept of force is conspicuously absent from our most advanced formulations of the basic laws. It doesn't appear in Schrödinger's equation, or in any reasonable formulation of quantum field theory, or in the foundations of general relativity. Astute observers commented on this trend to eliminate force even before the emergence of relativity and quantum mechanics.

In his 1895 *Dynamics*, the prominent physicist Peter G. Tait, who was a close friend and collaborator of Lord Kelvin and James Clerk Maxwell, wrote:

> In all methods and systems which involve the idea of force there is a leaven of artificiality. . . . There is no necessity for the introduction of the word "force" nor of the sense-suggested ideas on which it was originally based.[1]

Particularly striking, since it is so characteristic and so over-the-top, is what Bertrand Russell had to say in his 1925 popularization of relativity for serious intellectuals, *The ABC of Relativity:*

> If people were to learn to conceive the world in the new way, without the old notion of "force," it would alter not only their physical imagination, but probably also their morals and politics. . . . In the Newtonian theory of the solar system, the sun seems like a monarch whose behests the planets have to obey. In the Einsteinian world there is more individualism and less government than in the Newtonian.[2]

The fourteenth chapter of Russell's book is entitled "The Abolition of Force."

If $F = ma$ is formally empty, microscopically obscure, and maybe even morally suspect, what's the source of its undeniable power?

To track that source down, let's consider how the formula gets used.

A popular class of problems specifies a force and asks about the motion, or vice versa. These problems look like physics, but they are exercises in differential equations and geometry, thinly disguised. To make contact with physical reality, we have to make assertions about the forces that actually occur in the world. All kinds of assumptions get snuck in, often tacitly.

The zeroth law of motion, so basic to classical mechanics that Newton did not spell it out explicitly, is that mass is conserved. The mass of a body is supposed to be independent of its velocity and of any forces imposed on it; also total mass is neither created nor destroyed, but only redistributed, when bodies interact. Nowadays, of course, we know that none of that is quite true.

Newton's third law states that for every action there's an equal and opposite reaction. Also, we generally assume that forces do not depend on velocity. Neither of those assumptions is quite true either; for example, they fail for magnetic forces between charged particles.

When most textbooks come to discuss angular momentum, they introduce a fourth law, that forces between bodies are directed along the line that connects them. It is introduced in order to "prove" the conservation of angular momentum. But this fourth law isn't true at all for molecular forces.

Other assumptions get introduced when we bring in forces of constraint, and friction.

I won't belabor the point further. To anyone who reflects on it, it soon becomes clear that $F = ma$ by itself does not provide an algorithm for constructing the mechanics of the world. The equation is more like a common language, in which different useful insights about the mechanics of the world can be expressed. To put it another way, there is a whole culture involved in the interpretation of the symbols. When we learn mechanics, we have to see lots of worked examples to grasp properly what force really means. It is not just a matter of building up skill by practice; rather, we are imbibing a tacit culture of working assumptions. Failure to appreciate this is what got me in trouble.

The historical development of mechanics reflected a similar learning process. Isaac Newton scored his greatest and most complete success in planetary astronomy, when he discovered that a single force of quite a simple form dominates the story. His attempts to describe the mechanics of extended bodies and fluids in the second book of *The Principia*[3] were pathbreaking but not definitive, and he hardly touched the more practical side of mechanics. Later physicists and mathematicians—including notably Jean d'Alembert (constraint and contact forces), Charles Coulomb (friction), and Leonhard Euler (rigid, elastic, and fluid bodies)—made fundamental contributions to what we now comprehend in the culture of force.

Many of the insights embedded in the culture of force, as we've seen, aren't completely correct. Moreover, what we now think are more correct versions of the laws of physics won't fit into its language easily, if at all. The situation begs for two probing questions: How can this culture continue to flourish? Why did it emerge in the first place?

For the behavior of matter, we now have extremely complete and accurate laws that in principle cover the range of phenomena addressed in classical mechanics and, of course, much more. Quantum electrodynamics (QED) and quantum chromodynamics (QCD) provide the basic laws for building up material bodies and the nongravitational forces between them, and general relativity gives us a magnificent account of gravity. Looking down from this exalted vantage point, we can get a clear perspective on the territory and boundaries of the culture of force.

Compared to earlier ideas, the modern theory of matter, which really only emerged during the twentieth century, is much more specific and prescriptive. To put it plainly, you have much less freedom in interpreting the symbols. The equations of QED and QCD form a closed logical system: They inform you what bodies can be produced at the same time as they prescribe their behavior; they govern your measuring devices—and you, too!—thereby defining what questions are well posed physically; and they provide answers to such questions—or at least algorithms to arrive at the answers. (I'm well aware that QED + QCD is not a complete theory of nature, and that, in practice, we can't solve the equations very well.) Paradoxically, there is much less interpretation, less culture involved in the foundations of modern physics than in earlier, less complete syntheses. The equations really do speak for themselves: They are algorithmic.

By comparison to modern foundational physics, the culture of force is vaguely defined, limited in scope, and approximate. Nevertheless it survives the competition, and continues to flourish, for one overwhelmingly good reason: It is much easier to work with. We really do not want to be picking our way through a vast Hilbert space, regularizing and renormalizing ultraviolet divergences as we go, then analytically continuing Euclidean Green's functions defined by a limiting procedure, working to discover nuclei that clothe themselves with electrons to make atoms that bind together to make solids, all to describe the collision of two billiard balls. That would be lunacy similar in spirit to, but worse than, trying to do computer graphics from scratch, in machine code, without the benefit of an operating system. The analogy seems apt: Force is a flexible construct in a high-level language, which, by shielding us from irrelevant details, allows us to do elaborate applications relatively painlessly.

Why is it possible to encapsulate the complicated deep structure of matter? The answer is that matter ordinarily relaxes to a stable internal state, with high energetic or entropic barriers to excitation of all but a few degrees of freedom. We can focus our attention on those few effective degrees of freedom; the rest just supply the stage for the actors.

While force itself does not appear in the foundational equations of modern physics, energy and momentum certainly do, and force is very closely related to them: Roughly speaking, it's the space derivative of the former and the time derivative of the latter (and $F = ma$ just states the consistency of those definitions!). So the concept of force is not quite so far removed from modern foundations as Tait and Russell insinuate: It may be gratuitous, but it is not bizarre. Without changing the content of classical mechanics, we can cast it in Lagrangian terms, wherein force no longer appears as a primary concept. But that's really a technicality; the deeper questions remain: What aspects of fundamentals does the *culture* of force reflect? What approximations lead to it?

Some kind of approximate, truncated description of the dynamics of matter is both desirable and feasible because it is easier to use and focuses on the relevant. To explain the rough validity and origin of specific concepts and idealizations that constitute the culture of force, however, we must consider their detailed content. A proper answer, like the culture of force itself, must be both complicated and open-ended. The molecular explanation of friction is still very much a research topic, for example.

Here I conclude with some remarks on the psychological question, why force was—and usually still is—introduced in the foundations of mechanics, when from a logical point of view energy would serve at least equally well, and arguably better. The fact that changes in momentum—which correspond, by definition, to forces—are visible, whereas changes in energy often are not, is certainly a major factor. Another is that, as active participants in statics—for example, when we hold up a weight—we definitely feel we are doing something, even though no mechanical work is performed. Force is an abstraction of this sensory experience of exertion. D'Alembert's substitute, the virtual work done in response to small displacements, is harder to relate to. (Though ironically it is a sort of virtual work, continually made real, that explains our exertions. When we hold a weight steady, individual muscle fibers contract in response to feedback signals they get from spindles; the spindles sense small displacements, which must get compensated before they grow.[4]) Similar reasons may explain why Newton used force. A big part of the explanation for its continued use is no doubt (intellectual) inertia.

1. Peter G. Tait, *Dynamics* (London: Adam & Charles Black, 1895).

2. Bertrand Russell, *The ABC of Relativity*, 5th rev. ed. (London: Routledge, 1997).

3. Issac Newton, *The Principia*, trans. I. B. Cohen, A. Whitman (Berkeley: University of California Press, 1999).

4. Steven Vogel, *Prime Mover: A Natural History of Muscle* (New York: Norton, 2001), p. 79.

PETER GALISON

Einstein's Compass

FROM *SCIENTIFIC AMERICAN*

As a child, Albert Einstein was first exposed to—and captivated by—science when he was shown a magnetic compass. Peter Galison, a historian of science, studies a fascinating byway in Einstein's career, when he returned to the compass to resolve some of the mysteries of magnetism.

A t the beginning of 1915 Albert Einstein found himself engaging more and more in politics; he started to protest the militarism that had plunged Europe into a devastating war. That year also marked a significant change in the path of his long life in science. Collaborating with mathematician Marcel Grossman, Einstein was scrambling to learn all he could about a new kind of geometry, heretofore almost entirely unknown to physicists, that might aid him in characterizing the bending of spacetime. The stakes, he realized, were vast: Could special relativity be generalized into a theory of gravity? Could the Newtonian cosmos of distant inverse-square forces be scrapped in favor of one based on the equivalence of mass and energy with fields of curved space and time? In November 1915, after the most intense intellectual struggle of his life, Einstein was finally able to reveal general relativity to the world. His gargantuan effort was no less than a triumph of theory, reason, and abstraction.

Yet from the start and through much of that eventful year, Einstein had stepped back from the Platonic reaches of tensors and coordinate transformations to focus on bench experiments involving gluing quartz fibers to mirrors and pulsing electric currents through electromagnets. As he wrote to his best friend, Michele Besso, on February 12: "The experiment will soon be finished. . . . A wonderful experiment, too bad you can't see it. And how devious nature is, if one wants to approach it experimentally! I've gotten a longing for experiment in my old age." Working with Hendrik Lorentz's son-in-law, W. J. de Haas, Einstein undertook an experimental challenge that had stumped some of the most adept lab hands of all time—explaining the mechanism responsible for magnetism in iron.

The basic concept was simple. An electric current traveling in a loop makes an electromagnet. Einstein wondered whether magnetized iron might not also owe its capacity for magnetization to a similar phenomenon, as André Marie Ampère and his successors had long speculated. Einstein asked whether, at the atomic or molecular level, there were many such current loops all oriented in the same direction. If so, there might be just one kind of magnetism. Said he:

> Since [Hans Christian] Oersted discovered that magnetic effects are produced not only by permanent magnets but also by electrical currents, there may have been two seemingly independent mechanisms for the generation of the magnetic field. This state of affairs itself brought the need to fuse together two essentially different field-producing causes into a single one—to search for a single cause of the production of the magnetic field. In this way, shortly after Oersted's discovery, Ampère was led to his famous hypothesis of molecular currents which established magnetic phenomena as arising from charged molecular currents.[1]

Reducing two causations to one: here was quintessential Einstein. He had begun his work on special relativity with the assertion that the usual understanding of James Clerk Maxwell's equations must be very wrong, because it seemed as if there were two explanations for why current was produced when a wire coil approached a magnet. If the coil was moving and the magnet still, the standard story held that this was because the charge in the coil was moving (along with the wire) and so was pulled around the loop by the magnetic field. If the magnet was moving toward the coil, then, according to the conventional view, the growing magnetic field near the coil was produced by an electric field that drove charge around the coil. Einstein's special theory of relativity accounted for both phenomena by reassessing the meaning of space, time, and simultaneity.

In his 1907 principle of equivalence, Einstein had objected to the previously unchallenged claim that there were two kinds of mass—gravitational mass (responsible for the weight of a lead ball) and inertial mass (the resistance of a mass, say a lead ball, to acceleration, even far out in space). Instead Einstein stated that there was just one kind of mass. There was no way to distinguish the behavior of mass pressed to the floor of an accelerating rocket ship and that of mass pulled to the floor of a stationary room in a gravitational field.

So Einstein likewise believed deeply that there was but one kind of magnetism and that it was caused by the aligned orientation of tiny magnets—current loops formed by electrons as they raced around atomic nuclei. The question was: How could one test this idea?

Suppose that you are standing on a lazy Susan with a gyroscope in each hand, each with its axis pointing away from you and spinning clockwise from your point of view. The gyroscopes' angular momenta are oriented in opposite directions, so the system's total angular momentum adds to zero. Next, say you raise your hands above your head so the gyroscopes are now both pointing up. This means their angular momenta are both aimed in the same direction, so they sum to a nonzero value. But because the angular momentum in a closed system is conserved (stays the same), you begin to rotate on the lazy Susan, in this case to counter the angular momenta of the gyros.

Einstein imagined this scenario in miniature, inside an iron bar.

Suppose that an unmagnetized iron cylinder was suspended by a fine, flexible fiber and that suddenly a strong magnetic field was applied, enough to magnetize the cylinder by orienting all the little electron orbits. If he was correct, many of the little randomly oriented electron orbits would then be aligned. Their angular momenta would suddenly add instead of canceling. And again, just as the lazy Susan did, the cylinder would rotate to compensate. This was the notion behind the experiment. In time, amazingly, Einstein and de Haas succeeded in eliciting results from the remarkably delicate apparatus they built subsequently. But from where did this concept come, and why just then in 1915, amid the worst war and his own high-stakes struggle to define general relativity?

For an answer, one must look back to the period after Einstein's graduation from the Zurich Polytechnic in 1900, years during which he found it difficult to find gainful employment. Rejection letters piled up until mid-1902, when he finally received a very welcome job offer from the Bern Patent Office. Although Einstein had battled with one teacher after another during his school years, he admired and learned much from the head of the patent office, Friedrich Haller. Einstein learned to adhere strictly to Haller's injunction to "remain critically vigilant"—to view inventors' claims with skepticism.

Einstein loved machines and corresponded with other enthusiasts about them; he even built new ones in his apartment. Over the years he patented refrigerators, invented new electrical measurement devices and advised his friends about machinery. Indeed, his father and uncle had long run an electrotechnical business and had patented their own inventions. Sadly for us, nearly all of Einstein's patent evaluations were, by law, destroyed, but a few remain—in particular, those that made their way into court proceedings. That is because Einstein soon became one of the most esteemed technical authorities in the patent office and thus a much appreciated expert witness.

Herein lies the key to understanding Einstein's fascination with magnetism. In the early twentieth century the tried-and-true magnetic compass began to suffer difficulties. It worked poorly on new ships, which were becoming metallic and electrified, and functioned badly inside submarines or near the Earth's poles. And the standard compass was problematic in aircraft because its directional indicator led and lagged during turns.

Two companies took up the compass problem, one headed by American inventor and industrialist Elmer A. Sperry and the other by his German archrival, Hermann Hubertus Maria Anschütz-Kaempfe. The solution was to convert powered gyroscopes into compasses. Anschütz-Kaempfe cleverly built the casing of his gyroscope so that it would precess (slowly cycle its axial orientation) in such a way that its axis lined up with the rotational axis of the Earth. Soon afterward, Sperry produced a similar instrument. Anschütz-Kaempfe promptly sued for patent infringement. Sperry mounted the usual defense: he was merely following an older, preexisting idea.

In mid-1915 Einstein was called in to serve as an expert witness. His testimony showed, to the court's satisfaction, that the earlier gimbaled gyroscopes could not possibly have worked as compasses, because they were designed to move only within a very tight range inside their casings—a ship's slightest pitch and yaw would render them useless. Anschütz-Kaempfe won the case. Einstein went on to become sufficiently expert in gyrocompass technology to collect royalties for his work in this field for decades to come.

Einstein's royalties in the science of physics proved to be even greater, however: "I was led to the demonstration of the nature of the paramagnetic atom through technical reports I had prepared on the gyromagnetic compass."[2] He saw that just as the Earth's rotation oriented a gyrocompass, a cylinder of iron could be made to rotate by orienting all the little atomic gyroscopes inside it. The experiment turned out to be a spectacular success. Einstein and de Haas had demonstrated an effect so subtle that even the great James Clerk Maxwell had failed to discern it.

But this story has a twist. The two physicists showed excellent agreement between the theory (ferromagnetism caused by orbiting electrons) and their experiment. Unfortunately, their striking result soon came under attack—cautiously at first, then with growing insistence. It seemed that their measurement of magnetism per unit of angular momentum was off by a factor of two, a difference no one could adequately explain until much later, after the development of quantum mechanics and the concept of electron spin. It seems that Einstein's commitment to a particular theoretical model had cut two ways. On the one hand, it had given him real conviction about how to organize and conduct the experiment—specifically, where to look for the effect. Maxwell and others who had failed before had no feeling for the magnitude of the phenomenon. On the other hand, the theoretical model Einstein chose made it easy to accept an experimental answer when blackboard calculation and laboratory results agreed— despite the existence of many potentially interfering factors, which included such things as the effect of the Earth's magnetic field and the vagaries of the fragile lab apparatus itself.

The tale reminds me of one of Einstein's wonderful sayings: "No one but a theorist believes his theory; everyone puts faith in a laboratory result but the experimenter himself."

1. Albert Einstein and W. J. de Haas, "Experimenteller Nachweis der Ampèreschen Molekularströme," *Deutsche Physikalische Gesellschaft* 17 (1915), p. 152.

2. Albert Einstein to E. Meyerson, January 27, 1930, Einstein Archives Online.

WILLIAM J. BROAD

Will Compasses Point South?

FROM THE *NEW YORK TIMES*

As if there isn't enough to worry about, scientists are now observing signs of an impending reversal in the Earth's magnetic field, something that last happened 780,000 years ago. While it's not quite a doomsday scenario, William J. Broad previews the disruptions in store.

The collapse of the Earth's magnetic field, which both guards the planet and guides many of its creatures, appears to have started in earnest about one hundred and fifty years ago. The field's strength has waned 10 to 15 percent, and the deterioration has accelerated of late, increasing debate over whether it portends a reversal of the lines of magnetic force that normally envelop the Earth.

During a reversal, the main field weakens, almost vanishes, then reappears with opposite polarity. Afterward, compass needles that normally point north would point south, and during the thousands of years of transition, much in the heavens and Earth would go askew.

A reversal could knock out power grids, hurt astronauts, and satellites, widen atmospheric ozone holes, send polar auroras flashing to the equator and confuse birds, fish, and migratory animals that rely on the steadiness of the magnetic field as a navigation aid. But experts said the repercussions would fall short of catastrophic, despite a few proclamations of doom and sketchy evidence of past links between field reversals and species extinctions.

Although a total flip may be hundreds or thousands of years away, the rapid decline in magnetic strength is already damaging satellites.

In June 2004 the European Space Agency approved the world's largest effort at tracking the field's shifts. A trio of new satellites, called Swarm, are to monitor the collapsing field with far greater precision than before and help scientists forecast its prospective state.

"We want to get some idea of how this would evolve in the near future, just like people trying to predict the weather," said Dr. Gauthier Hulot, a French geophysicist working on the satellite plan. "I'm personally quite convinced we should be able to work out the first predictions by the end of the mission."

The discipline is one of a number—like high-energy physics and aspects of space science—where Europeans have recently come from behind to seize the initiative, dismaying some American experts.

No matter what the new findings, the public has no reason to panic, scientists say. Even if a flip is imminent, it might take two thousand years to mature. The last one took place 780,000 years ago, when *Homo erectus* was still learning how to make stone tools.

Some experts suggest a reversal is overdue. "The fact that it's dropping so rapidly gives you pause," said Dr. John A. Tarduno, a professor of geophysics at the University of Rochester. "It looks like things we see in computer models of a reversal."

In an interview, Dr. Tarduno put the odds of an impending flip at more likely than not, adding that some of his colleagues were placing informal bets on the possibility but realized they would probably be long gone by the time the picture clarified.

Deep inside the Earth, the magnetic field arises as the fluid core oozes with hot currents of molten iron and this mechanical energy gets converted into electromagnetism. It is known as the geodynamo. In a car's generator, the same principle turns mechanical energy into electricity.

No one knows precisely why the field periodically reverses, but scientists say the responsibility probably lies with changes in the turbulent flows of molten iron, which they envision as similar to the churning gases that make up the clouds of Jupiter.

In theory, a reversal could have major effects because over the ages many aspects of nature and society have come to rely on the field's steadiness.

When baby loggerhead turtles embark on an eight-thousand-mile trek around the Atlantic, they use invisible magnetic clues to check their bearings. So do salmon and whales, honeybees and homing pigeons, and frogs and Zambian mole rats, scientists have found.

On a planetary scale, the magnetic field helps shield the Earth from solar winds and storms of deadly particles. Its so-called magnetosphere extends out thirty-seven thousand miles from Earth's sunlit side and much farther behind the planet, forming a cometlike tail.

Among other things, the field's collapse, scientists say, could let in bursts of radiation, causing a variety of disruptions.

Dr. Charles H. Jackman, an atmospheric scientist at NASA's Goddard Space Flight Center in Greenbelt, Maryland, has worked with European colleagues on a computer model that mimics the repercussions. A weak field, they reported in December 2003, could let solar storms pummel the atmosphere with enough radiation to destroy significant amounts of the ozone that protects the Earth from harmful ultraviolet light.

Ultraviolet radiation, the short, invisible rays from the sun, can harm some life forms, depress crop yields and raise cancer rates, causing skin cancer and cataracts in humans. Dr. Jackman said that the ozone damage from any one solar storm could heal naturally in two to three years but that the protective layer would stay vulnerable to new bursts of radiation as long as the Earth's magnetic field remained weak.

"It would be significant" in terms of planetary repercussions, he said in an interview, "but not catastrophic." High levels of ultraviolet radiation would spread down from polar regions as far south as Florida.

Like many of the Earth's invisible rhythms, the field reversals are typically slow, taking anywhere from five thousand to seven thousand years to complete.

Strong evidence of their reality first emerged in the 1950s and 1960s when scientists towing magnetic sensors behind ships found that the rocky seabed exhibited odd stripes of magnetization.

It turned out that continuous flows of seabed lava became alternately magnetized over the ages as the polarities of the Earth's field switched one way, then the other. The seabed acted like a huge tape recorder, and the same proved true of the layered deposits of old volcanoes on land.

How did the rocky memories form? Molten lava proved to hold tiny mineral grains that acted like innumerable compasses, or miniature magnets, freely aligning themselves with the contemporary field. But as the lava cooled, the tiny compasses froze in place, immobile even if the field shifted. Experts called it paleomagnetism and found that the tiny compasses were often made of magnetite, a naturally magnetic mineral.

Paleomagnetic studies showed that the Earth's field reversed every half million years or so, but in a fairly random way and with early patterns more chaotic. During the age of dinosaurs, for instance, no flips occurred for roughly thirty-five million years.

As scientists began to understand the importance of reversals in the planet's history, they examined the fossil record for evidence of damage to life. In 1971 Dr. James D. Hays of the Lamont-Doherty Earth Observatory of Columbia University noted a strong correlation between recent flips and species extinctions of tiny marine creatures known as radiolarians. "The evidence," he wrote in *The Bulletin of the Geological Society of America,* "is strongly suggestive that magnetic reversals either directly or indirectly exert a selective force."

But no consensus ever formed on how the flips might have doomed some creatures and spared others, and some experts faulted the correlations as statistically insignificant.

Meanwhile, starting in the late 1970s, scientists began to find wide evidence that many animals relied on the Earth's magnetic field for navigation. Dr. Joseph L. Kirschvink of the California Institute of Technology, discovered such reliance in bees, pigeons, bacteria, salmon, whales, and newts, among other animals. The magnetic sense, he found, usually relies on tiny crystals of magnetite—the same mineral that gets immobilized in cooling lava.

Investigators looking into the origin of the reversals got new clues in 1995 when scientists at the Los Alamos National Laboratory and the University of California at Los Angeles succeeded in making the first computer simulation of the geodynamo in action, including field reversals.

Dr. Gary A. Glatzmaier, who was one of the Los Alamos scientists, said it showed that the Earth's solid inner core resisted the flipping because the field there could not change as rapidly as it did in the fluid outer core. "The reversal starts with a small region that gets larger," he said in an interview. "Most of the time they die away, but other times they continue to grow." To date, the simulations of millions of years have produced more than a dozen flips.

The current collapse drew wide scientific attention on April 11, 2002, when *Nature*, the British journal, published a major paper that detailed its growing weakness. Dr. Hulot and colleagues at the Institut de Physique du Globe de Paris, where he works, as well as the Danish Space Research Institute, called the large drop remarkable.

They found it by comparing readings made in 1979 and 1980 by the American Magsat satellite with measurements by the Danish Oersted satellite, launched in 1999 and still operating. In particular, Dr. Hulot and his team discovered a north polar region and a spot below South Africa where the magnetism is growing extremely weak.

The finding drew wide attention because the magnetic anomalies seemed consistent with what the computer simulations identified as the possible beginnings of a flip.

"We postulate," Dr. Hulot and his coauthors wrote, that the new evidence reflects how "the geodynamo operates before reversing."

In an interview, he said that the field's southern spot was 30 percent weaker than elsewhere and that some satellites passing over it had already suffered electronic malfunctions when highly charged particles from the sun were able to penetrate the weakened magnetic shield.

In March 2003 *The Core,* a Hollywood film, gave a wildly exaggerated portrayal of what would happen if the field vanished. People with pacemakers fall dead. Pigeons fly into people and windows. And the planet, a scientist warns, will fry in a year.

Dr. Tarduno said that practical effects on things like satellites and the ozone layer would be the same no matter whether the field reversed or simply weakened and bounced back. A major collapse of the Earth's magnetic shield, he added, could let speeding particles penetrate deeper into the atmosphere to widely knock out power grids, as solar storms do occasionally.

The consensus among biologists seems to be that the reversals are slow enough, and the Earth's creatures resilient enough, that most would learn to adapt. They note the lack of correlations in the fossil record between flips and mass extinctions.

Dr. Kenneth J. Lohmann, a biologist at the University of North Carolina who has pioneered magnetic navigation studies in loggerhead turtles, said if the field became weak enough "there would be problems for the turtles." His research suggests they use it not only for a general sense of direction but as a precise map of their location.

To better understand the current collapse, the European Space Agency plans to launch three satellites in 2009. The spacecraft, flying in polar orbits a few hundred miles up, are to map its intricacies until perhaps 2015.

Dr. Hulot said scientists would combine the satellite data with computer simulations to make not only distant forecasts but possible warnings of current hazards. Among the possible solutions would be to increase satellite shielding.

"It will be interesting to see what's going to happen in that South Atlantic anomaly," he said. "If you want to keep satellites flying, you want to know if the situation is going to deteriorate."

K. C. COLE

Seeking Life as We Know It

FROM THE *LOS ANGELES TIMES*

With recent missions to Mars pointing to the historic presence of water, and thus the tantalizing possibility of life, K. C. Cole investigates whether water is actually a necessary requirement.

Albert Einstein once famously wondered whether God had a choice in how he created the universe. His unanswered question drives physics to this day.

The same question could be asked about the biological universe—especially now that the rover *Opportunity* has found signs of ancient standing water on Mars.

NASA's search for alien life is based on the strategy "follow the water," and for obvious reasons. The only life we know is built on a scaffolding of carbon that floats in bags of water. Bacteria or brontosaurus, we're all made from the same basic recipe.

But did life have a choice? Could it have evolved from entirely different ingredients? In looking for water-based life in worlds beyond, are we making the mistake of peering into a mirror?

Why not life in ethanol? suggested Cornell University's Roald Hoffmann, a Nobel laureate in chemistry. Or ammonia?

"Now life in liquid ammonia, that would be colorful," said Hoffmann, explaining that metals can dissolve in ammonia, "giving bright blue solutions."

And why does the scaffolding have to be carbon?

Why not silicon, its neighbor on the periodic table of elements?

"We're so dumb about what life is because we only have one example," said astrobiologist Chris McKay, of NASA Ames Research Center at Moffett Field, near the Bay Area city of Mountain View. "It may be true that we sail through the universe and everything we find is carbon and water, but I would hesitate to conclude that based on the one example we have."

As a practical matter, NASA's strategy of following the water makes good sense.

"We don't know how to do anything better," McKay said. "We're too stupid to look for things if we don't know what they are."

At $820 million, the twin rover missions have to look at what's most likely. "If you had to bet, what would you bet on?" asked Stanford chemist Richard Zare.

Still, one has to wonder what else might be out there.

The search is complicated by the fact that scientists aren't even sure what life is exactly. Bizarre new species are discovered on Earth all the time in the most unlikely places.

"We even have trouble understanding what's alive and what's dead," Zare said. "People still wonder what a virus is."

All life as we know it is spun from carbon-based threads swimming in water solutions. Both carbon and water have unique—some say magical—properties. Indeed, physics and chemistry strongly suggest that life might not have had a choice.

Water is the most eccentric of liquids. "It's this elusive, magical, mystery molecule," said James Garvin, lead scientist for the Mars exploration program at NASA headquarters in Washington.

On the face of it, water seems a rather silly molecule—two hydrogen atoms attached to an oxygen atom in a way that looks like the head of Mickey Mouse. Even children know its chemical formula:

H_2O. But the bonds it forms with itself and other molecules are anything but ordinary.

Atoms normally bond by sharing the negatively charged electrons that buzz around their positively charged nuclei, like people sharing popcorn at a movie. In water, the oxygen shares one electron with each of its hydrogens, leaving four extras. These clump together as "lone pairs" that can grab onto other molecules like prehensile feet.

At the same time, the two positive hydrogen nuclei stick out the other side like arms. The "feet" of one water molecule grab the "arms" of the other, forming abnormally strong networks. Where one water molecule goes, the others tend to follow. Thus, water can climb tall trees—hand over foot, as it were—in defiance of gravity, carrying nutrients from the soil to the leaves.

Chemists say they would expect water to be a gas at room temperature because it's made up of just a few light atoms. But the strong bonds make the molecules stick together in a liquid form.

Luckily, the bonds aren't so sticky that they form a viscous gel—something that Boston University physicist Eugene Stanley initially found perplexing. Water flows freely, he and others discovered, because water molecules stick to each other only briefly, let go, grab another partner—whirling an ever-changing cast of partners around in a molecular square dance. The upshot is that water stays watery over a remarkable range of temperatures (32 to 212 degrees Fahrenheit, to be exact).

This is a liquid bonanza for life, which seems to need some form of fluid to transport things from place to place. In solids, molecules stick together and can't go much of anywhere. In gases, the molecules don't get close enough to interact.

Water's unbalanced geometry—positive charges on one side, negative on the other—also gives it a distinctively schizophrenic personality (although chemists, like psychiatrists, prefer the term bipolar). This makes it an excellent solvent.

One side of a molecule grabs on to negative charges; the other side grabs the positive. This pulls most things apart, so water can dissolve almost anything. (If things didn't dissolve, they'd sink to the bottom, or rise to the top—not good for a free flow of chemical reactions.)

Why doesn't life just disintegrate altogether in water then? While water is one of the most strongly bipolar molecules, it is not the most reactive—meaning it can make things fall apart (dissolve) without changing their composition (react). So the parts can be endlessly rearranged.

And as it turns out, the few things water doesn't dissolve are equally important in assembling life's building blocks. Water hates fat. "It won't dissolve a spot of grease on my nice silk tie," Stanley said.

Water herds these hydrophobic (water-hating) and hydrophilic (water-loving) molecules into structures such as cells. The hydrophobes point away from each other, while the hydrophiles look inward. "It's like circling the wagons," McKay said.

Water, in other words, gives living things outsides and insides. The hostile outside is kept at bay, while inside, the proteins behind nearly all of life's mechanisms go about their business.

"You have three thousand proteins, minimally, in every cell," said University of Massachusetts biologist Lynn Margulis, "and every reaction requires water. Everything else is negotiable."

What's the water doing with the proteins exactly? "Everything," Margulis said. "It's like a loom that you can do the weaving in. It's the matrix that's holding things in place. Nothing can go on without it."

The magical molecule does a whole lot more: For example, it absorbs heat slowly, and holds on to it for a long time. This stabilizes temperatures not only in the oceans but also inside living things—which, lest we forget, are made mainly of water.

Finally, water expands when it freezes, contrary to nearly every other substance known. That's why ice floats, allowing it to form an insulating blanket on lakes and ponds for life beneath. Without it, fish would freeze before they hit the grocer's shelves.

Of course, it's hard to ignore one obvious reason life may depend on water. Hydrogen is the most abundant element in the universe.

Helium is the second, but it's inert—so standoffish it doesn't bond with other atoms at all. Oxygen comes third. Maybe life is made of water simply because it's there.

But some otherwise habitable worlds just don't have water. Are they out of luck?

Not necessarily. "Water's a wonderful molecule," McKay said, "but there are other wonderful molecules."

Ethanol, or grain alcohol, would probably work, concurred UCLA chemist Ken Houk. Proteins and nucleic acids are soluble in ethanol. But the liquid is rare in nature because the chemistry needed to produce it is complicated.

In contrast, water "is the easiest fluid to make," Garvin said.

As for ammonia (used in smelling salts), it's scarce on Earth, but "you could easily have an ocean of ammonia," Houk said. In fact, scientists speculate that Saturn's moon Titan could have such an ocean. Life could certainly exist at the cold temperatures at which ammonia is liquid (between minus 28 degrees and minus 108 degrees on Earth). Like water, ammonia is polar, and an excellent solvent.

Even if water does turn out to be the beverage of choice for quenching life's insatiable thirst, does that mean carbon has to be in the mix too?

Many scientists think it does.

"I feel more strongly about carbon than about water," said David Des Marais, an astrobiologist at NASA Ames Research Center.

Again, there's an abundance argument. Carbon is the fourth-most common element. And life grabs the ingredients at hand.

Carbon also has unique properties that allow it to form long chains and rings easily. Think of carbon as a small atom with four Velcro (actually electronic) attachment points. One, two, or three of these can form links with other atoms, giving carbon enormous versatility. Almost anything can find a way to attach. So carbon just naturally makes the kinds of complex molecules life needs.

Like water, carbon is a Goldilocks substance: It forms strong, stable bonds, but not so strong that those bonds can't break off and attach to something else. "You have this kind of texture," Margulis said, "a range of properties that change in very subtle ways."

Carbon's closest competitor, silicon, is not so subtle. Sitting right below carbon on the periodic table of elements, it also has four attachment points, but it's heavier and has different chemical properties.

It can make long chains if you add oxygen, for example. But then everything it touches turns to stone. "It locks on to things, and folks, it's over," Zare said. "It's very hard to break the bonds. It's like rigor mortis." So virtually any attempt at metabolism as we know it would produce something solid.

Solid silicon compounds are already familiar—as rocks, glass, gels, bricks, and, of course, medical implants.

Life seems to have ignored silicon, except here and there as structural material in rice, grasses, and microscopic algae. How ironic, Hoffmann noted, that the silicon worlds we build ourselves (computers, electronics) now dominate our lives. "This is silicon's revenge!"

If there were such a thing as silicon life, it would have to be built on an entirely different biological model. It probably would be stiff—unable to breathe, for example, as we do.

"You'd have to give up not just carbon but the whole pattern," McKay said. "We live as bags of liquid. A better model [for silicon life] is more like computers, a rigid life form that gets its energy from some electrochemical means directly."

Just because we do our chemistry on the inside, he said, doesn't mean all life does. Silicon life might do its chemistry on the surface.

But if silicon life appeared on ancient Earth along with carbon life, as some speculate (rather wildly) that it might have, it wouldn't stand a chance from an evolutionary perspective.

"You might be able to make living things out of different materials," said UCLA planetary scientist David Paige. "But I'm comfortable with the idea that the life we are is the best that we could do given the constraints of our environment and the laws of physics and chemistry."

Those laws of physics and chemistry apply to the entire universe, so life elsewhere, Paige speculates, might well look familiar. "If we find a planet that's covered with water, the life forms are likely to look like fish, because there's a good reason fish look like fish and dolphins and submarines."

Of course, life can't spring from carbon and water alone.

At a minimum, life also needs some form of energy—the kind we use from the Sun, or the heat of radioactive decay from deep inside the Earth, or tidal friction that comes from being a large moon (like Titan) orbiting a large planet.

Life, at its essence, is a mechanism for turning energy into order.

Many purely physical processes do that as well: Gravity herds stars into galaxies. The late Columbia University physicist Gerald Feinberg and New York University biochemist Robert Shapiro speculated that what they called "physical life" could exist in solid hydrogen, in neutron stars, even in interstellar clouds, living on the energy of radiation. This "radiant life" would consist of individual beings they called "radiobes."

"It may be difficult to think of such systems of being alive," they acknowledged in an article included in the collection *Extraterrestrials: Where Are They?* But our own biochemistry—based on proteins and nucleic acids—does little "to convey the wonders, such as elephants and Sequoia trees, that ultimately arise from it."

Would we recognize these alternative life forms if we saw them? Probably not.

"Our imagination is biased by what we're able to see," Paige said. "We can't be as clever as the universe. So we have to be careful."

One of the mistakes of the 1976 *Viking* missions to Mars, Paige said, was looking for life that was "too lifelike." Life, for example, that eats familiar kinds of food, thrives in similar environments.

Since that time, scientists have discovered bizarre new biological worlds of so-called extremophiles on Earth, thriving in places where life was thought to be impossible—such as boiling-hot vents at the bottom of the ocean, shut off from sunlight, subsisting on hydrogen sulfide.

These life forms (giant tube worms, for example) came as a complete surprise. Now, many scientists believe they may be our earliest ancestors.

More surprises are certainly in store. "We still don't understand how life works," Houk said. "It's utterly miraculous. Even though it's sitting there and staring us in the face, we don't understand it."

Dennis Overbye

All of a Sudden, the Neighborhood Looks a Lot Friendlier

FROM THE *NEW YORK TIMES*

> *Contemplating the possibility that we might soon detect life else-*
> *where in the galaxy, Dennis Overbye pauses to celebrate all the uni-*
> *verse still has to teach us.*

Like most New Yorkers, I have real estate fever. Even though I hate moving, I can't travel anywhere without wondering what it would be like to live there. I can't walk down a street in Oaxaca or the East Village without window shopping for apartments and evaluating the restaurant scene and the availability of playgrounds.

It doesn't stop there. Roll a sleeping bag out under the sky in a place like Mesa Verde, seven thousand feet up in the Colorado Rockies, on a summer evening and you will wake up at midnight with your nose in the Milky Way. There are roughly two hundred billion stars in the Milky Way and somewhere around that number of galaxies in the observable universe. Surely there must be some prime real estate out there somewhere, perhaps with neighbors.

These are hopeful times for those inclined to view the universe as generous and nurturing, who like to think the prospects of life elsewhere are good. As of early September 2004 the official count of verified extrasolar planets stood at 111, according to a list kept by the International Astronomical Union at www.ciw.edu/boss/IAU/div3/wgesp/planets.shtml.

But the numbers change weekly. What began as a trickle of discoveries of strange unlivable planets a decade ago has become a torrent of new ones looking friendlier and friendlier to life.

One recent discovery orbits Gliese 436, a dim star slightly less than half the mass of the Sun, known as a red dwarf. It was announced in September 2004, that Gliese 436 has a planet only about twenty times the mass of Earth, about the size of Neptune in our own system.

Circling its parent star every 2.6 days at a distance of about 2.6 million miles, this new planet would be unlivable, like all the other much bigger exoplanets already discovered—except for one circumstance that can't be ruled out or in, yet. And here is where the chain of speculation and dreams begins.

The planet is so close to its star, astronomers calculate, that tidal forces have locked it into keeping the same face toward the star, the way the Moon keeps the same face toward Earth as it rotates. That means that one side of the planet is perpetually blowtorched while the other is in an eternal deep freeze. But in the twilight zone, between the two sides, where Gliese 436 looms like an orange clown face forever on the horizon, there is a thread of a chance, some astronomers say, that the temperature could be in the Goldilocks range, neither too hot nor too cold, of liquid water—the sine qua non of Life As We Know It.

I have begun to imagine a kind of ring world on Gliese 436 b, as this planet is now unpoetically known, a dusky narrow scrum of hungry green, a sort of Alaskan summer bog strip crawling with ingenious critters hemmed between ice and desert, building their narrow little cities straddling a canal full of extraterrestrial bass, with windows pointed sunward, full of clever mirrors.

It's interesting to ponder the psychology and cosmology of creatures who live in perpetual twilight. Is it always dawn or always sunset on Gliese 436 b? Would they know about stars?

Of course this is a long shot. Forget about cities and life. A few alien bacteria in a mud puddle someplace would change science. Even the possible existence of liquid water someplace would have astrobiologists and science writers doing handstands, but that thread hangs on the nature of the planet's atmosphere. If it's too thick, heat from the illuminated side would be conducted around to the dark side and the whole place would be stifling.

If the atmosphere is thin or nonexistent, astronomers concede, some more grudgingly than others, it's not impossible that some piece of the twilight zone could remain lukewarm. But nobody really knows what Gliese 436 b is like, whether it is made of rock and iron, like the Earth, or ice and snow like Neptune, nor what the dynamics of its atmosphere might be, whether, for example, water would evaporate away from the light and fall as snow on the cold side.

And it's worse than that. Red dwarfs like Gliese 436 are prone to giant flares and sunspots, so its planet and anything on it would have to endure variations in sunshine as well as radiation showers.

It's way too soon for sober folks to get excited about the possibility of life at Gliese 436 or any other particular putative dust speck in the sky. But odds are increasingly in favor of the idea that one of these days, some newly discovered planet will have the "Goldilocks" properties that will make it a good bet for an enterprising real estate agent.

The newly discovered planets were all detected indirectly, mainly by looking for the wobbles their gravity induces in the motions of their stars, and so the race is on at observatories around the world to get pictures. Because they create the biggest wobble, the first exoplanets were gaseous giants like Jupiter orbiting close to their stars, too hot and dense for any life that we know about. But as they have looked longer and more sensitively, astronomers are beginning to find friendlier-looking systems.

Gliese 436 b is only one of three new planets recently discovered that are significantly smaller, about the size of Neptune. One of them, known as 55 Cancri e, is part of a four-planet system that is the closest thing astronomers have found yet to our own solar system.

These discoveries have moved the game into a whole new realm, unexpectedly quickly, said Dr. Geoff Marcy, an astronomer and longtime planet hunter at the University of California, Berkeley.

When he and his collaborator, Dr. Paul Butler of the Carnegie Institution, were starting out in the early 1990s, Dr. Marcy said, "people literally looked down at their shoes in embarrassment when we told them that we were trying to detect planets." Hunting for planets around other stars, he said, was considered akin to searching for UFOs.

Now it's mainstream astronomy and a growing chunk of NASA's budget.

The next big milestone, astronomers say, will be the detection of Earth-size planets, although that will require going to space. NASA has planned a series of missions, including *Kepler,* scheduled for 2007, the *Space Interferometry Mission,* scheduled for 2009, and a pair of spacecraft known as *Terrestrial Planet Finders* about a decade from now to identify and study habitable planets.

Closer to home, the *Spirit* and *Opportunity* rovers seem to have confirmed the presence of water sometime in the past on Mars. In September 2004, Scientists from the University of Colorado have concluded from analyzing data from the *Mars Global Surveyor* and the *Mars Odyssey* spacecraft that *Opportunity*'s landing site had once been an ocean as big as the Baltic Sea.

Scientists think there are oceans under the ice on Jupiter's moon Europa, and hydrocarbons raining down onto the slushy surface of Saturn's moon Titan, into which the *Huygens* probe will descend this winter.

Nobody knows if there ever was or is life in any of these places. Scientists have found life everywhere on Earth that they have looked, from the driest and the coldest to the hottest and wettest of places, but the odds of life off Earth are incomputable under the best conditions we can imagine or hope for.

Scientists don't know how life originated on Earth, after all. No less a light than Francis Crick, codiscoverer of the structure of DNA, suggested in an article in 1973 in *Icarus,* a planetary science journal, along with the chemist Dr. Leslie Orgel, that extraterrestrials could have spread life throughout the stars in the form of spores or genetic material encapsulated in small meteorites, an idea that goes all the way back to the Greek philosopher Anaxagoras.

Nor are scientists absolutely sure they even know how to recognize life that is too far removed from the one example we know—ourselves and our kin—based on the chemistry of DNA. In the search for extraterrestrial life, as Dr. Jill Tarter, a radio astronomer at the SETI Institute in Mountain View, Calif., once pointed out, "Number two is still the most important number in the process."

The question of life has recently come to occupy a central position not just in the space program but in physics and cosmology, traditionally the most bloodless of sports, requiring a depressingly vast and abstract perspective on the cosmos. Cosmologists have found to their astonishment that life and the universe are strangely and deeply connected. Life As We Know It seems to depend on the miraculous and improbable juggling of the numerical values of a few atomic and astronomical constants—like the relative masses of elementary particles or the strength of the "dark energy" accelerating the universe outward. Twiddle these control knobs on Nature's console a little bit and the galaxies evaporate before stars and life have a chance to evolve, or atoms fall apart.

What this means is being debated fiercely these days by some of the smartest physicists in the world. Do we live in a lucky universe? Or are there zillions of different possible universes and we live in one with laws that happen to be conducive to life, the way that fish live on a planet that is warm enough to have liquid water?

"We live where we can live," say the proponents of the latter view. But others, following the example of Albert Einstein, say that we don't know anything about physics yet and that we if we work hard enough, we will understand why God had to make the universe the way it is.

Now I am not generally known for my optimism. I think the world is getting worse and more dangerous, despite the onward march of science. And I don't expect to hear the end of the debate about the alleged fine-tuning of the laws of nature, which seems to me to be as much about philosophy and personality as it is about science.

What seems indisputably clear is that our knowledge of the universe is dwarfed by our ignorance.

So I find myself bullish on the universe, so to speak. Almost everything is still possible. I think it is likely that astronomers will find Earth-like planets in my lifetime and probably long before anybody lands on Mars. We might live to see it as a little dot in the newspapers or on television, and to start thinking about sending a space probe there, knowing it will be a voyage that would last generations.

We might find out where it is that we can live. As it happens, my family and I just finished moving into a new apartment. By the time we get an answer back, I might be ready to move again.

How Will the Universe End?

FROM *SLATE*

It used to be enough to speculate that the world will end in fire or ice, but now cosmologists are entertaining a third possibility—a runaway expansion that rips the universe apart. Tracking the latest thinking on the future of the universe, and humanity's place in it, Jim Holt offers a tongue-in-cheek look at some of the more fanciful speculations of modern cosmology.

One of my favorite moments in Woody Allen's film *Annie Hall* is when Alvie Singer (Allen's alter ego) is shown having an existential crisis as a little boy. His mother summons a psychiatrist, one Dr. Flicker, to find out what's wrong.

"Why are you depressed, Alvie?" Dr. Flicker asks.

"The universe is expanding," Alvie says. "The universe is everything, and if it's expanding, some day it will break apart and that will be the end of everything."

"Why is that your business?" interrupts his mother. Turning to the psychiatrist, she announces, "He's stopped doing his homework!"

"What's the point?" Alvie says.

"What has the universe got to do with it!" his mother shouts. "You're here in Brooklyn! Brooklyn is not expanding!"

Dr. Flicker jumps in: "It won't be expanding for billions of years, Alvie, and we've got to enjoy ourselves while we're here, eh? Ha ha ha." (Cut to a view of the Singer house, which happens to be under the Coney Island roller coaster.)

I used to take Dr. Flicker's side in this matter. How silly to despond about the end of everything! After all, the cosmos was born only around thirteen billion years ago, when the Big Bang happened, and parts of it will remain hospitable to our descendants for a good hundred billion years, even as the whole thing continues to spread out.

A half-dozen years ago, however, astronomers peering through their telescopes began to notice something rather alarming. The expansion of the universe, their observations indicated, was not proceeding at the stately, ever-slowing pace that Einstein's equations had predicted. Instead, it was speeding up. Some "dark energy" was evidently pushing against gravity, sending galaxies hurtling away from one another at a runaway rate. New measurements earlier this year confirmed this strange finding. On July 22, 2003, the *New York Times* ran an ominous headline: "Astronomers Report Evidence of 'Dark Energy' Splitting the Universe." David Letterman found this so disturbing that he mentioned it several consecutive nights in his *Late Show* monologue, wondering why the *Times* buried the story on Page A-13.

Until recently, the ultimate destiny of the universe looked a little more hopeful—or remote. Back around the middle of the last century, cosmologists figured out that there were two possible fates for the universe. Either it would continue to expand forever, getting very cold and very dark as the stars winked out one by one, the black holes evaporated, and all material structures disintegrated into an increasingly dilute sea of elementary particles: the Big Chill. Or it would eventually stop expanding and collapse back upon itself in a fiery, all-annihilating implosion: the Big Crunch.

Which of these two scenarios would come to pass depended on one crucial thing: how much stuff there was in the universe. So, at least, said Einstein's theory of general relativity. Stuff—matter and energy—creates gravity. And, as every undergraduate physics major will tell you, gravity sucks. It tends to draw things together. With enough stuff, and hence enough gravity, the expansion of the universe would eventually be arrested and reversed. With too little stuff, the gravity would merely slow the expansion, which would go on forever. So, to determine how the universe would ultimately expire, cosmologists thought that all they had to do was to weigh it. And preliminary estimates—taking account of the visible galaxies, the so-called dark matter, and even the possible mass of the little neutrinos that swarm though it all—suggested that the universe had only enough weight to slow the expansion, not to turn it around.

Now, as cosmic fates go, the Big Chill might not seem a whole lot better than the Big Crunch. In the first, the temperature goes to absolute zero; in the second, it goes to infinity. Extinction by fire or by ice—what's to choose? Yet a few imaginative scientists, haunted, like Woody Allen, by visions of the end of the universe, came up with formulations of how our distant descendants might manage to go on enjoying life forever, despite these unpleasant conditions. In the Big Chill scenario, they could have an infinity of slower and slower experiences, with lots of sleep in between. In the Big Crunch scenario, they could have an infinity of faster and faster experiences in the run-up to the final implosion. Either way, the progress of civilization would be unlimited. No cause for existential gloom.

So, Letterman had reason to be upset by the dark energy news. It spells inescapable doom for intelligent life in the far, far future. No matter where you are located, the rest of the universe would eventually be receding from you at the speed of light, slipping forever beyond the horizon of knowability. Meanwhile, the shrinking region of space still accessible to you will fill up with a kind of insidious radiation that would eventually choke off information processing—and with it, the very possibility of thought. We seem to be headed not for a Big Crunch or a Big Chill but something far nastier: a Big Crackup. "All our knowledge, civilization and culture are destined to be forgotten," one prominent cosmologist has declared to the press. It looks as if little Alvie Singer was right after all. The universe is going to "break apart," and that will indeed mean the end of everything—even Brooklyn.

Hearing this news made me think of the inscription that someone once said should be on all churches: *important if true.* Applied to cosmology—the study of the universe as a whole—that is a big "if." Cosmic speculations that make it into the newspapers should often be taken with a pinch of salt. A few years ago, some astronomers from Johns Hopkins made headlines by announcing that the cosmos was turquoise; two months later they made headlines again by announcing that, no, it was actually beige. This may be a frivolous example, but even in graver matters—like the fate of the universe—cosmologists tend to reverse themselves every decade or so. As one of them once told me, cosmology is not really a science at all since you can't do experiments with the universe. It's more like a detective story. Even the term that is sometimes applied to theorizing about the end of the universe, "eschatology" (from the Greek word for "furthest"), is borrowed from theology.

Before I was going to start worrying about the extinction of absolutely everything in some inconceivably distant epoch, I thought it would be a good idea to talk to a few leading cosmologists. Just how certain were they that the cosmos was undergoing a disastrous run-away expansion? Was intelligent life really doomed to perish as a result? How could they, as scientists, talk about the ultimate future of "civilization" and "consciousness" with a straight face?

It seemed natural to start with Freeman Dyson, an English-born physicist who has been at the Institute for Advanced Study in Princeton since the 1940s. Dyson is one of the founding fathers of cosmic escha-tology, which he concedes is a "faintly disreputable" subject. He is also a fierce optimist about the far future, one who envisions "a universe growing without limit in richness and complexity, a universe of life surviving forever and making itself known to its neighbors across unimaginable gulfs of space and time." In 1979 he wrote a paper called "Time Without End," in which he used the laws of physics to show

how humanity could flourish eternally in a slowly expanding universe, even as the stars died and the chill became absolute. The trick is to match your metabolism to the falling temperature, thinking your thoughts ever more slowly and hibernating for longer and longer periods while extraneous information is dumped into the void as waste heat. In this way, Dyson calculated, a complex society could go on perpetually with a finite energy reserve, one equivalent to a mere eight hours of sunlight.

The day I went to see Dyson, it was raining in Princeton. It took me a half hour to walk from the train station to the Institute for Advanced Study, which sits by a pond in five hundred acres of woods. The institute is a serene, otherworldly place. There are no students to distract the eminent scientists and scholars in residence from pursuing their intellectual fancies. Dyson's office is in the same building where Einstein spent the last decades of his career fruitlessly searching for a unified theory of physics. An elfin, courtly man with deep-set eyes and a hawklike nose, Dyson frequently lapsed into silence or emitted snuffles of amusement. I started by asking him whether the evidence that the universe was caught up in an accelerated expansion had blighted his hopes for the future of civilization.

"Not necessarily," he said. "It's a completely open question whether this acceleration will continue forever or whether it will peter out after a while. There are several theories of what kind of cosmic field might be causing it and no observations to determine which of them is right. If it's caused by the so-called dark energy of empty space, then the expansion will keep speeding up forever, which is bad news as far as life is concerned. But if it's caused by some other kind of force field—which, out of ignorance, we label 'quintessence'—then the expansion might well slow down as we go into the future. Some quintessence theories even say that the universe will eventually stop expanding altogether and collapse. Of course, that, too, would be unfortunate for civilization since nothing would survive the Big Crunch."

Well, then, I said, let's stick with the optimistic scenario. Suppose the acceleration does turn out to be temporary and the future universe settles into a nice cruise-control expansion. What could our descendants possibly look like a trillion trillion trillion years from now, when the stars have disappeared and the universe is dark and freezing and so diffuse that it's practically empty? What will they be made of?

"The most plausible answer," Dyson said, "is that conscious life will take the form of interstellar dust clouds." He was alluding to the kind of inorganic life forms imagined by the late astronomer Sir Fred Hoyle in his 1957 science fiction novel, *The Black Cloud.* "An ever-expanding network of charged dust particles, communicating by electromagnetic forces, has all the complexity necessary for thinking an infinite number of novel thoughts."

How, I objected, can we really imagine such a wispy thing, spread out over billions of light-years of space, being conscious?

"Well," he said, "how do you imagine a couple of kilograms of protoplasm in someone's skull being conscious? We have no idea how that works either."

Practically next door to Dyson at the institute is the office of Ed Witten, a gangly, fifty-ish fellow who is widely regarded as the smartest physicist of his generation, if not the living incarnation of Einstein. Witten is one of the prime movers behind superstring theory, which, if its hairy math is ever sorted out, may well furnish the Theory of Everything that physicists have long been after. He has an unnerving ability to shuffle complicated equations in his head without ever writing anything down, and he speaks in a hushed, soft voice. Earlier this year, Witten was quoted in the press calling the discovery of the runaway expansion of the universe "an extremely uncomfortable result." Why, I wondered, did he see it that way? Was it simply inconvenient for theoretical reasons? Or did he worry about its implications for the destiny of the cosmos? When I asked him, he agonized for a moment before responding, "Both."

Yet Witten, too, thought there was a good chance that the runaway expansion would be only temporary, as some of the quintessence theories predicted, rather than permanent, as the dark-energy hypothesis implied. "The quintessence theories are nicer, and I hope they're right," he told me. If the acceleration does indeed relax to zero, and the Big Crackup is averted, could civilization go on forever? Witten was unsure. One cause for concern was the possibility that protons will eventually decay, resulting in the dissolution of all matter within another, oh, 10^{33} years or so. Freeman Dyson had scoffed at this when I talked with him, pointing out that no one had ever observed a proton decaying, but he insisted that intelligent beings could persist even if atoms fell to pieces, by reembodying themselves in "plasma clouds"—swarms of electrons and positrons. I mentioned this to Witten. "Did Dyson really say that?" he exclaimed. "Good. Because I think protons probably do decay."

BACK AT THE PRINCETON RAILROAD STATION after visiting Ed Witten and Freeman Dyson, waiting for the train to New York and munching on a vile "veggie" sandwich that I had picked up at the convenience store across the parking lot, I pondered proton decay and Dyson's scenario for eternal life. How would his sentient Black Clouds, be they made up of cosmic dust or of electron-positron plasma, while away the eons in an utterly freezing and dark universe? What passions would engross their infinite number of ever-slowing thoughts? After all (as Alvie Singer's alter ego once observed), eternity is a long time, especially toward the end. Maybe they would play games of cosmic chess, in which each move took trillions of years. But even at that rate they would run through every possible game of chess in a mere $10^{10^{70}}$) years—long before the final decay of the burnt-out cinders of the stars. What then? Would they come around to George Bernard Shaw's conclusion (reached by him at the age of ninety-two) that the prospect of personal immortality was an "unimaginable horror"? Or would they feel that, subjectively at least, time was passing quickly enough? After all, as Fran Lebowitz pointed out, once you've reached the age of fifty, Christmas seems to come every three months.

It was almost with a sense of relief that I spoke to Lawrence Krauss a few days later. Krauss, a boyish fellow in his late forties who teaches at Case Western Reserve in Cleveland, is one of the physicists who guessed on purely theoretical grounds, even before the astronomical data came in, that the cosmos might be undergoing a runaway expansion. "We appear to be living in the worst of all possible universes," Krauss told me, clearly relishing the note of anti-Leibnizian pessimism he struck. "If the runaway expansion keeps going, our knowledge will actually decrease as time passes. The rest of the universe will be literally disappearing before our very eyes surprisingly soon—in the next ten or twenty billion years. And life is doomed—even Freeman Dyson accepts that. But the good news is that we can't *prove* we're living in the worst of all possible universes. No finite set of data will ever enable us to predict the fate of the cosmos with certainty. And, in fact, that doesn't really matter. Because, unlike Freeman, I think that we're doomed even if the runaway phase turns out to be only temporary."

What about Dyson's vision of a civilization of sentient dust clouds living forever in an expanding universe, entertaining an infinite number of thoughts on a finite store of energy? "It turns out, basically for mathematical reasons, that there's no way you can have an infinite number of thoughts unless you do a lot of hibernating," Krauss said. "You sleep for longer and longer periods, waking up for short intervals to think—sort of like an old physicist. But what's going to wake you up? I have a teenage daughter, and I know that if I didn't wake her up, she'd sleep forever. The Black Cloud would need an alarm clock that would wake it up an infinite number of times on a finite amount of energy. When a colleague and I pointed this out, Dyson came up with a cute alarm clock that could actually do this, but then we argued that this alarm clock would eventually fall apart because of quantum mechanics."

So, regardless of the fate of the cosmos, things look pretty hopeless for intelligent life in the long run. But I should remember, Krauss said, that the long run is a very long time. He told me about a meeting he attended at the Vatican a few years back on the future of the universe: "There were about fifteen people, theologians, a few cosmologists, some biologists. The idea was to find common ground, but after three days it was clear that we had nothing to say to one another. When theologians talk about the 'long term,' raising questions about resurrection and such, they're really thinking about the short term. We weren't even on the same plane. When you talk about 10^{50} years, the theologians' eyes glaze over. I told them that it was important that they listen to what I had to say—theology, if it's relevant, has to be consistent with science. At the same time I was thinking, 'It doesn't matter what *you* have to say, because whatever theology has to say is irrelevant to science.' "

At least one cosmologist I knew of would be quite happy to absorb theology into physics, especially when it came to talking about the end of the universe. That's Frank Tipler, a professor at Tulane University in New Orleans. In 1994 Tipler published a strangely ingenious book called *The Physics of Immortality*, in which he argued that the Big Crunch would be the happiest possible ending for the cosmos. The final moments before universal annihilation would release an infinite amount of energy, Tipler reasoned, and that could drive an infinite amount of computation, which would produce an infinite number of thoughts—a subjective eternity. Everyone who ever existed would be "resurrected" in an orgy of virtual reality, which would correspond pretty neatly to what religious believers have in mind when they talk about heaven. Thus, while the physical cosmos would come to an abrupt end in the Big Crunch, the mental cosmos would go on forever.

Was Tipler's blissful eschatological scenario—which he called "the Omega Point"—spoiled by the news that the cosmos seemed to be caught up in a runaway expansion? He certainly didn't think so when I talked to him. "The universe has *no choice* but to expand to a maximum size and then contract to a final singularity," he exclaimed in his thick Southern drawl. (He's a native of Alabama and a self-described "redneck.") Any other cosmic finale, he said, would violate a certain law of quantum mechanics called "unitarity." Moreover, "the known laws of physics *require* that intelligent life persist until the end of time and gain control of the universe." When I mentioned that Freeman Dyson (among others) could not see why this should be so, Tipler shouted in exasperation, "Ah went up to Princeton last November and ah *tode* him the argument! Ah *tode* him!" Then he told me, too. It was long and complicated, but the nub of it was that intelligent beings must be present at the end to sort of massage the Big Crunch in a certain way so that it would not violate another law of quantum mechanics, the "Beckenstein bound." So, our eternal survival is built into the very logic of the cosmos. "If the laws of PHEE-ysics are with us," he roared, "who can be against us?"

Tipler's idea of an infinite frolic just before the Big Crunch was seductive to me—more so, at least, than Dyson's vision of a community of increasingly dilute Black Clouds staving off the cold in an eternal Big Chill. But if the universe is in a runaway expansion, both are pipe dreams. The only way to survive in the long run is to get the hell out. Yet how do you escape a dying universe if—as little Alvie Singer pointed out—the universe is everything?

A man who claims to see an answer to this question is Michio Kaku. A theoretical physicist at City College in New York, Kaku looks and talks a bit like the character Sulu on *Star Trek*. (He can be seen in the recent Michael Apted film about great scientists, *Me and Isaac Newton.*) He is not the least bit worried about the fate of this universe. "If your ship is sinking," he said to me, "why not get a lifeboat and leave?" We earthlings can't do this just yet, Kaku observed. That is because we are a mere Type 1 civilization, able to marshal the energy only of a single planet. But eventually, assuming a reasonable rate of economic growth and technological progress, we will graduate to being a Type 2 civilization, commanding the energy of a star, and thence to being a Type 3 civilization, able to summon the energy of an entire galaxy. Then space-time itself will be our plaything. We'll have the power to open up a "wormhole" through which we can slip into a brand-new universe.

"Of course," Kaku added, "it may take as long as one hundred thousand years for such a Type 3 civilization to develop, but the universe won't start getting really cold for *trillions* of years." There is one other thing that the beings in such a civilization will need, Kaku stressed to me: a unified theory of physics, one that would show them how to stabilize the wormhole so it doesn't disappear before they can make their escape. The closest thing we have to that now, superstring theory, is so difficult that no one (with the possible exception of Ed Witten) knows how to get it to work. Kaku wasn't the least bit gloomy that the universe might be dying. "In fact," he said, "I'm in a state of exhilaration, because this would force us, really force us, to crack superstring theory. People say, 'What has superstring theory done for *me* lately? Has it given me better cable TV reception?' What I tell them is that superstring theory—or whatever the final, unified theory of physics turns out to be—could be our one and only hope for surviving the death of this universe."

Although other cosmologists were rudely dismissive of Kaku's lifeboat scenario—"a good prop for a science fiction story," said one; "somewhat more fantastical than most of *Star Trek*," remarked another—it sounded good to me. But then I started thinking. To become a Type 3 civilization, one powerful enough to engineer a stable wormhole leading to a new universe, we would have to gain control of our entire galaxy. That means colonizing something like a billion habitable planets. But if this is what the future is going to look like, then almost all the intelligent observers who will ever exist will live in one of these billion colonies. So, how come we find ourselves sitting on the home planet at the very beginning of the process? The odds against being in such an unusual situation—the very earliest people, the equivalent of Adam and Eve—are a billion to one.

MY VAGUE QUALM about the unlikeliness of Kaku's lifeboat theory was considerably sharpened when I talked to J. Richard Gott III, an astrophysicist at Princeton University. Gott is known for making bold quantitative predictions about the longevity of things—from Broadway shows like *Cats* to America's space program to intelligent life in the universe. He bases these predictions on what he calls the Copernican Principle, which says, in essence: *You're not special.* "If life in the universe is going to last a long time, why do we find ourselves living when we do, only thirteen billion years after the beginning?" Gott said to me, speaking in an improbable Tennessee accent whose register occasionally leapt up an octave, like Don Knotts'. "And it is a disturbing fact that we as a species have only been around for two hundred thousand years. If there are going to be many intelligent species descended from us flourishing in epochs far in the future, then why are we so lucky to be the first?" Doing a quick back-of-the-envelope calculation, Gott determined that it was 95 percent likely that humanity would last more than 5,100 years but would die out before 7.8 million years (a longevity that, coincidentally, is quite similar to that of other mammal species, which tend to go extinct around two million years after appearing). Gott was not inclined to speculate on what might do us in—biological warfare? asteroid collision? nearby supernova? sheer boredom with existence? But he did leave me feeling that the runaway expansion of our universe, if real, was the least of our worries.

Despite the pessimistic tenor of Gott's line of thought, he was positively chirpy in conversation. In fact, all the cosmologists I had spoken to so far had a certain mirthfulness about them when discussing eschatological matters—even Lawrence Krauss, the one who talked about this being the worst of all possible universes. ("Eschatology—it's a great word," Krauss said. "I had never heard of it until I discovered I was doing it.") Was no one made melancholy or irritable by the prospect of our universe decaying into nothingness? I thought of Steven Weinberg, the Nobel laureate in physics who, in his 1977 book about the birth of the universe, *The First Three Minutes*, glumly observed, "The more the universe seems comprehensible, the more it also seems pointless." It was Weinberg's pessimistic conclusion in that book—he wrote that civilization faced cosmic extinction from either endless cold or unbearable heat—that had inspired Freeman Dyson to come up with his scenario for eternal life in an expanding cosmos.

I called Weinberg at the University of Texas, where he teaches. "So, you want to hear what old grumpy has to say, eh?" he growled in a deep voice. He began with a dazzling theoretical exposition that led up to a point I had heard before: No one really knows what's causing the current runaway expansion or whether it will continue forever. The most natural assumption, he added, was that it would. But he wasn't really worried about the existential implications. "For me and you and everyone else around today, the universe will be over in less than 10^2 years," he said. In his peculiarly sardonic way, Weinberg seemed as jolly as all the other cosmologists. "The universe will come to an end, and that may be tragic, but it also provides its fill of comedy. Postmodernists and social constructivists, Republicans and socialists and clergymen of all creeds—they're all an endless source of amusement."

IT WAS TIME to tally up the eschatological results. The cosmos has three possible fates: Big Crunch (eventual collapse), Big Chill (expansion forever at a steady rate), or Big Crackup (expansion forever at an accelerating rate). Humanity, too, has three possible fates: eternal flourishing, endless stagnation, or ultimate extinction. And judging from all the distinguished cosmologists who weighed in with opinions, every combination from Column A and Column B was theoretically open. We could flourish eternally in virtual reality at the Big Crunch or as expanding black clouds in the Big Chill. We could escape the Big Crunch/Chill/Crackup by wormholing our way into a fresh universe. We could face ultimate extinction by being incinerated by the Big Crunch or by being isolated and choked off by the Big Crackup. We could be doomed to endless stagnation—thinking the same patterns of thoughts over and over again, or perhaps sleeping forever because of a faulty alarm clock—in the Big Chill. One distinguished physicist I spoke to, Andrei Linde of Stanford University, even said that we could not rule out the possibility of their being something *after* the Big Crunch. For all of the fascinating theories and scenarios they spin out, practitioners of cosmic eschatology are in a position very much like that of Hollywood studio heads: *Nobody knows anything.*

Still, little Alvy Singer is in good company in being soul-sick over the fate of the cosmos, however vaguely it is described. At the end of the nineteenth century, figures like Swinburne and Henry Adams expressed similar anguish at what then seemed to be the certain heat-death of the universe from entropy. In 1903 Bertrand Russell described his "unyielding despair" at the thought that "all the labors of the ages, all the devotion, all the inspiration, all the noonday brightness of human genius, are destined to extinction in the vast death of the solar system, and that the whole temple of Man's achievement must inevitably be buried beneath the debris of a universe in ruins." Yet a few decades later, he declared such effusions of cosmic angst to be "nonsense," perhaps an effect of "bad digestion."

Why should we want the universe to last forever, anyway? Look— either the universe has a purpose or it doesn't. If it doesn't, then it is absurd. If it does have a purpose, then there are two possibilities: Either this purpose is eventually achieved, or it is never achieved. If it is never achieved, then the universe is futile. But if it is eventually achieved, then any further existence of the universe is pointless. So, no matter how you slice it, an eternal universe is either (a) absurd, (b) futile, or (c) eventually pointless.

Despite this cast-iron logic, some thinkers believe that the longer the universe goes on, the better it is, ethically speaking. As John Leslie, a cosmological philosopher at the University of Guelph in Canada, told me, "This is true simply on utilitarian grounds: The more intelligent happy beings in the future, the merrier." Philosophers of a more pessimistic kidney, like Schopenhauer, have taken precisely the opposite view: Life is, on the whole, so miserable that a cold and dead universe is preferable to one teeming with conscious beings.

If the current runaway expansion of the cosmos really does portend that our infinitesimal flicker of civilization will be followed by an eternity of bleak emptiness, then that shouldn't make life now any less worth living, should it? It may be true that nothing we do in AD 2004 will matter when the burnt-out cinder of our sun is finally swallowed by a galactic black hole in a trillion trillion years. But by the same token, nothing that will happen in a trillion trillion years matters to us now. In particular (as the philosopher Thomas Nagel has observed), *it does not matter now that in a trillion trillion years nothing we do now will matter.*

Then what is the point of cosmology? It's not going to cure cancer or solve our energy problems or give us a better sex life, obviously enough. Still, it is bracing to realize that we live in the first generation in the history of humanity that might be able to answer the question, How will the universe end? "It amazes me," Lawrence Krauss said, "that, sitting in a place on the edge of nowhere in a not especially interesting time in the history of the universe, we can, on the basis of simple laws of physics, draw conclusions about the future of life and the cosmos," he said. "That's something we should relish, regardless of whether we're here for a long time or not."

So, remember the advice offered by Monty Python in their classic "Galaxy Song." When life gets you down, the song says, and you're feeling very small and insecure, turn your mind to the cosmic sublimity of the ever-expanding universe—"because there's bugger-all down here on Earth."

NATALIE ANGIER

Scientist at Work: Jacqueline Barton

FROM THE *NEW YORK TIMES*

Despite recent gains, men still outnumber women in the sciences by a wide margin, so an outstanding woman scientist like Jacqueline Barton is in great demand, as a chemist and a role model. Natalie Angier gets a glimpse of Dr. Barton's busy life and remarkable mind.

Like many a bio-minded scientist in this gilded age of the genome, Dr. Jacqueline K. Barton, a professor of chemistry at the California Institute of Technology, considers the double helix so aesthetically pleasing that she keeps a chubby, cheery model of it smack in the middle of her office table.

The three-foot-high sculpturette of some twenty base pairs of DNA—a tiny fraction of the three billion nucleic building blocks that make up the human genome—has a sort of Henry Moore bulbosity to it, coupled with the graceful, emergent torque of the Venus de Milo. All very sinusoidal and maternal: just what one would expect of the famed Molecule of Life and the mother of all cookbooks.

Yet as Dr. Barton speaks of her studies about how DNA might keep itself fit and fiery by shooting electrons up and down its span, the familiar object on her table begins to evoke a very different image: that of a lightning bolt, a crackling zigzag of opportunity. Far from being a couch-bound custodian of information, it seems, the molecule of life is the original live wire.

"What many people don't realize is how dynamic the structure of DNA is," said Dr. Barton, her fingers fluttering lightly over the model. "The base pairs are always moving and vibrating, electrons are migrating, holes are opening up and closing through the center of the DNA." It's like a cocktail party or a kindergarten class, she said. "Nothing stays still for more than a femtosecond here or a millisecond there."

Yet for all the squirm and spark, DNA, the ultimate source of information on how cells and bodies should behave, remains a remarkably stable presence throughout one's terrestrial tenure, and even beyond. It's because the molecule is so substantive and slow to degrade, said Dr. Barton, "that you can still look at dinosaur DNA."

Dr. Barton, fifty-one, and her colleagues are seeking to understand just how the double helix manages to be at once so twitchy and so reliable, capable of constant interchange with tens of thousands of proteins and other small characters in the cell, hammered at by blistering chemicals, ultraviolet rays, and corrosive free radicals, and expected to split and split and split again, spawning numberless generations of daughter DNA molecules in the course of cell division; and all the while still staying sane and functional and relatively error-free.

Dr. Barton proposes that the DNA molecule polices itself electronically, periodically delivering a flow of charged particles from Point A to Point B to check for mutant, misplaced bases that might be skulking in the corridors. If the electrons proceed unimpeded, she suggests, all is well. But if there is a kink in the sequence, the smallest sign of a nascent mutation, the flow would short-circuit. That break would in turn sound an alarm, alerting the cell's DNA repair crew to fix the mess now, or at least sometime before lunch.

The work is part of Dr. Barton's larger exploration of the electrical properties of DNA, and how the trafficking of charged particles in and around its zigs, swags, and crevasses may help mold the contours of the master macro-molecule or set its timbre twanging.

Dr. Barton is a heavy-metal fan. She and her lab mates have created metal-studded molecular probes that allow them to generate electrons when and where they want them, and to follow the migration of the particles as they ping their way through DNA. By taking an electron's-eye view of the double helix, she hopes to help solve a mystery of molecular biology: how all the tens of thousands of tiny proteins responsible for servicing the genome and carrying out its encoded directives find their correct genetic targets along an otherwise forbidding spaghetti of chromosomal coils.

On a practical note, Dr. Barton has designed tiny metal-based probes to test the integrity of a given DNA sample as quickly and cheaply as possible. A company that she founded in San Diego, GeneOhm Sciences Inc., is now working to translate some of her ideas into a simple laboratory assay that could detect disease mutations in a patient's DNA by simply running a current through it and seeing if anything trips it up.

Such a procedure would be much simpler than using biochemical methods to spell out, or sequence, thousands of DNA subunits in search of the occasional miscreant. Her work is also attracting attention from experts in the fashionable field of nanotechnology, the science of the extremely tiny, who would dearly love to exploit nature's microcircuitry in devices of their own.

Dr. Barton has long been a high-wattage figure on the chemistry circuit. She is a MacArthur fellow, a member of the National Academy of Sciences and a winner of laurels from the National Science Foundation, the American Chemical Society and the American Philosophical Society, to name a few.

Like the molecule she studies, Dr. Barton is a braid of contradictions: animated and composed, optimistic and pragmatic, genial and driven. "If the house were to burn down, if things were going badly in the lab, if a paper wasn't accepted by the right journal, she wouldn't get angry or frustrated or wear it on her sleeve, the way I might," said Dr. Peter Dervan, fifty-eight, her husband and also a professor of chemistry at Caltech.

Dr. Barton is compact, dark-eyed, dark-haired, and, true to her New York roots, she likes dressing in black. She walks briskly, as nearly all scientists do, but with a decided bounce in her step. "What can I say?" said Dr. Robert G. Shulman, an emeritus professor of molecular biophysics and biochemistry at Yale. "We were all jealous of her husband when they got married."

As one of the very few women ranking as a full professor at Caltech, and in the chemistry trade generally, Dr. Barton has perhaps inevitably become a role model to young women in science. In March 2004, for example, she participated in the first Rosalind Franklin International Lecture Program at the Imperial College in Britain, devised "to expose young scientists" to that wondrous sight, "internationally recognized women scientists."

Dr. Shana Kelley, an assistant professor of chemistry at Boston College, said, "I tell people she's found out how to do it all, that you don't have to choose between career or family, and that's what I want for myself."

"I might have had more doubts if I hadn't watched her," she added.

Dr. Anna Marie Pyle, a professor of molecular biophysics and biochemistry at Yale who studied under Dr. Barton, said: "Jackie conveys to her students the importance of creativity and imagination in science. Women need to have somebody express confidence in their creative abilities, not just their ability to work harder."

The good life has its grindstones and metronomes, though. The couple's time is so tightly scheduled, admitted Dr. Dervan, "that I can tell you where we'll be to the day, to the hour, for the next twelve months."

When they are at work, they work, and when they are at home, they are parents. (They have two children—Elizabeth, thirteen, and Andrew, twenty-one, Dr. Dervan's son from his first marriage.)

"People assume that because Peter and I are both chemists, we talk all the time about chemistry," said Dr. Barton. "In fact, we almost never talk shop. If we go out to dinner together, we talk about Elizabeth."

They travel often for work, but they are zealously efficient in those jaunts. Dr. Barton has been known to fly back and forth to Europe in a single day, alighting abroad just long enough to deliver a two-hour talk. "Who wants to sleep in a hotel alone," she said, "when you can sleep at home with your family?"

One of the few things that gets her ever so wryly ruffled is the public's attitude toward her beloved discipline, chemistry. "People hear the word 'chemical' and they automatically think it's bad," she said. "Even my daughter, when she was in kindergarten, came home and said she'd learned that there were too many chemicals in the world. I told her, Elizabeth, we're all chemicals! Everything is made of chemicals!"

Dr. Barton is also skeptical whenever laypeople tell her they "flunked chemistry" in high school. "Not everybody could possibly have flunked chemistry," she said. "Surely someone somewhere got a B."

For that matter, she said, why do people so readily confess to their complete ignorance of chemistry and other science, even to the point of sounding boastful? "People are perfectly willing to say to me at a party, I haven't the foggiest idea what you're talking about, which they would never do if we were discussing current events," she said. "Why aren't they embarrassed? Why don't they think, Gee, maybe this is something I ought to know a little bit about?" Fear is no excuse. For all its daunting reputation, she said: "Science really isn't that hard. I don't think it's more difficult than anything else."

Born and raised in New York City, she attended the Riverdale Country School for Girls. Back then, she said, "young girls didn't take chemistry."

The first chemistry course she took was in college, at Barnard. It became her major; she graduated summa cum laude. She earned her Ph.D. in inorganic chemistry from Columbia, and did a postdoctoral fellowship at Bell Labs and Yale. Dr. Barton then took a position as assistant professor of chemistry at Hunter College in New York.

She soon moved to Columbia, where her research using metal ions to study DNA began attracting considerable attention, from her future husband, among others. She and Dr. Dervan knew each other for years professionally before they began dating, and when Caltech sought to woo Dr. Barton from Columbia, she told the recruiters of the potentially confounding fact that she was in love with a member of their faculty. "Their response was, well, that's further incentive for you to come here," Dr. Dervan said.

Soon after moving to Caltech, Dr. Barton gave birth to her daughter, and within short order learned of her MacArthur "genius" award. "That came at a good time for me," she said. "I was wondering if people might think, oh, she's a mother now, is she going to start slowing down? The MacArthur gave me the reassurance that I was probably okay, and that I could go on to do interesting stuff."

So long as it doesn't require an overnight stay.

Jennifer Couzin

Aging Research's Family Feud

FROM *SCIENCE*

It's no longer a secret that science is not simply a dispassionate search for truth, but a competitive, sometimes cutthroat, race to new discoveries—and the rewards that go with them. Jennifer Couzin reports on the conflict between two prominent researchers studying the genetic causes of aging, a controversy made all the more emotional by the fact that one was the longtime mentor of the other.

At thirty-four, David Sinclair is a rising star. His spacious ninth-floor office at Harvard Medical School boasts a panoramic Boston view. His rapidly growing lab pulled off the feat of publishing in both *Nature* and *Science* in 2003, and it made headlines around the world with a study of the possible antiaging properties of a molecule found in red wine. In a typical day, he fields calls from a couple practicing a radical diet to extend life span, and from an actor hunting for antiaging pills and the chance to invest in Sinclair's new company.

There is, however, another side to this glossy picture of success. Sinclair is engaged in a tense and very public battle with his mentor, a renowned scientist based across the Charles River at the Massachusetts Institute of Technology (MIT). Leonard Guarente, fifty-one, is an undisputed leader in the field of aging, an author of major discoveries about genes that prolong life. For four years, Sinclair all but lived in Guarente's lab as a postdoctoral fellow, and the two grew extremely close.

Thanks in part to a glowing recommendation from Guarente, Sinclair nabbed a tenure-track spot on Harvard's faculty in late 1999. He then made clear that his old professor's pet theories weren't off limits. At a meeting at Cold Spring Harbor Laboratory in late 2002, Sinclair surprised Guarente by challenging him on how a key gene Guarente discovered extends life in yeast. That sparked a bitter dispute that crescendoed in the winter of 2004, when the pair published dueling papers.

Researchers who study aging are finding the quarrel both intellectually provocative and a lively source of gossip. And the reverberations extend well beyond that community. At its core, the argument involves one of the hottest topics in longevity research: how cutting calories may increase life span and how its effects can be translated into antiaging therapies. But the dispute, which remains unresolved, involves molecular biology so intricate that many scientists are uncertain how to assess it. It's also not clear how it applies to other organisms, such as mammals.

Guarente and Sinclair are also rivals in business. Both have bold ideas for translating yeast studies into ways to stall mammalian aging and treat age-related diseases. Sinclair recently announced plans to launch a biotechnology company that he says will compete directly with one cofounded by his mentor. "They're doing exactly what we're doing, and it's a race," says Sinclair, clearly relishing the prospect.

The two have spoken little since their Cold Spring Harbor falling-out. Some who know the pair say the clash is unsurprising; despite the generation gap, Guarente and Sinclair share many traits common among successful scientists. Both are deeply ambitious, relentlessly competitive, and supremely self-confident. Both savor the limelight. Both love science and show little interest in other pursuits.

Still, they've retained something of the parent-child relationship that can shape interactions between senior researchers and their students. Like a father dismayed when his son joins a punk rock band, and then dismayed further when it attracts devoted fans and favorable reviews, Guarente exhibits a mix of pride, anger, and disappointment when the conversation turns to his former postdoc. At the same time, Guarente confesses that Sinclair's choices aren't altogether startling. "There's a side of me that identifies with him," he says. "The young Lenny Guarente was not all that different."

A NATIVE OF MASSACHUSETTS, Guarente's academic life has revolved around two of the most high-powered institutions in the country. He attended MIT as an undergraduate—where biology at first "felt squishy to me"—and completed graduate and postgraduate work at Harvard. Then he moved two subway stops back up Massachusetts Avenue and settled again into MIT, where he has remained ever since.

With tenure under his belt at thirty-four, Guarente began thinking beyond the mainstream assignments to which he had gravitated early on, like studies of gene regulation. In the early 1990s two of his graduate students, Brian Kennedy and Nick Austriaco, sat down with him to discuss a project they wanted to pursue: dissecting the causes of aging.

At the time, aging was considered fringe science, a topic few reputable researchers would touch. But "Lenny likes a challenge," says Kennedy, now at the University of Washington, Seattle. "He said, 'You've got a year to learn something.' "

The clock ticking, Kennedy and Austriaco focused on yeast, a single-celled organism that lends itself to laboratory manipulation. They began hunting for mutant yeast cells that lived abnormally long life spans, measured by the number of daughter cells they produce. (A mother cell will typically produce a daughter every one to four hours; an average cell generates about twenty daughters.) In those early days, when Kennedy and Austriaco were testing hundreds of strains, they organized round-the-clock vigils to track their yeast cells; one of them was always there, gazing at the cells under a microscope and delicately counting off the daughters.

It quickly became obvious that some strains lived unusually long, which piqued Guarente's interest. He hovered nearby, checking in with the students two or three times a day and grilling them on what they'd found. The news was good: About one in one thousand strains both produced an abundance of daughter cells and survived well under the stress of a chilly refrigerator, underscoring known links between longevity and stress tolerance. Of these, one mutant caught their attention. It was sterile—unable to mate with other yeast cells—and lived 50 percent longer than normal.

Guarente canceled golf games with Kennedy to spend more time in the lab. Gradually, the group homed in on a trio of genes called *SIR* genes; deleting them seemed to shorten a yeast cell's life.

It was around this time, in early 1996, that twenty-six-year-old David Sinclair boarded a plane in his hometown of Sydney, Australia, and flew halfway around the world to Cambridge. "Before I even knew I got the [postdoctoral] fellowship, I said to Lenny, 'I'll take out a loan, I'll sell my car' " to come to the lab, says Sinclair. His wife-to-be, a German biologist, was moving to Australia on a fellowship of her own, but Sinclair couldn't pass up the opportunity to study with Guarente, whom he idolized. Still, he knew that the science Guarente favored was then heretical: "The idea that you could use yeast to study human aging was a joke."

Just as heresy had not deterred Guarente, it didn't stop Sinclair. He already knew yeast: As a graduate student at the University of New South Wales in Sydney, Sinclair had studied genetic regulation in yeast—he and Guarente first met after a genetics conference in Australia—and Sinclair's dissertation was the thickest in the lab. And he had already earned a reputation for pushing limits. He racked up traffic violations in his red sports car, regularly skating close to losing his license and once having it confiscated altogether. "You're allowed to get twelve points, and at one stage he had fourteen," recalls Geoff Kornfeld, the lab manager in Ian Dawes's lab at New South Wales, where Sinclair studied.

The atmosphere in Guarente's lab when Sinclair joined was electric. Other students were building on the yeast work by Kennedy and

Austriaco, and they pinpointed a specific *SIR* gene, called *SIR2*, that could dramatically extend the cells' life span. Extra copies of *SIR2* also enabled worms to live longer.

Furthermore, Guarente's lab had found that activity of the protein generated by *SIR2*, called SIR2p, was dependent on an enzyme called NAD. All cells carry NAD, which helps govern metabolism. SIR2p, it appeared, could sense the metabolic state of a yeast cell.

This was significant because it dovetailed with what scientists had observed for decades in animals: Curtailing calories and altering metabolism dramatically extends life span. Biologists theorize that, in the face of scarce resources, organisms trade off reproduction, which is hindered by calorie restriction, for survival. No one knows, though, precisely how a meager environment slows aging.

Guarente turned the connections over in his mind. Like calorie restriction, *SIR2* lengthened life span. Its connection with NAD and metabolism suggested something more: Maybe it was a gene that could explain why cutting calories slowed aging. Furthermore, when manipulated, perhaps *SIR2* could mimic the antiaging effects of calorie restriction without a near-starvation diet.

SIR2 wasn't the only project stirring excitement in Guarente's lab. A graduate student, David Lombard, had just cloned the mouse gene for Werner syndrome, a rare disease that mimics accelerated aging. Along with postdoc Robert Marciniak, Lombard was trying to understand how the Werner's protein behaved in mouse cells. "There were ideas and debates flying through the air constantly," says Brad Johnson of the University of Pennsylvania in Philadelphia, who was then a postdoc in the lab. Guarente, never one to coddle his students, pushed for results.

Sinclair proved to be a brilliant and prolific researcher, often the first to arrive, at 8:00 a.m., and the last to leave, at 12:30 a.m., running to catch the final subway train of the night. He quickly became a favorite of Guarente. But many lab members began regarding him warily, especially after the 1997 publication of a *Cell* paper by Sinclair and Guarente. The paper reported that buildup of ribosomal DNA, a kind of repetitive DNA sequence, in yeast cells caused them to age. Although Sinclair had conducted the experiments himself, Johnson had also proposed similar studies. When lab members learned that Johnson was not a coauthor on the paper, they began guarding their work more closely.

Sinclair has a ready reply. He says that when he first thought of the *Cell* experiments, he was concerned that others might accuse him of poaching the ideas. So, to prove that the concept was his, he wrote and mailed himself a letter describing the experiments before they'd been done. Sinclair still has that letter, its seal intact.

Sinclair agrees, though, that he was unpopular in Guarente's lab, but he explains it differently. "Lenny didn't hide his favoritism" for his Australian postdoc, says Sinclair.

GUARENTE'S ARMS ARE SORE from shoveling after a December blizzard dumped nearly two feet of snow on Boston, but that doesn't stop him from gesticulating to underline his side of the story. "This has run me through so many emotions, some of which I didn't know I had," he says of the falling-out with Sinclair.

Some facts are not in dispute: Calorie restriction extends life in nearly every species tested so far. In yeast, restricting calories boosts activity of *SIR2* proteins, and extra SIR2p slows aging. The critical question on which Guarente and Sinclair disagree is how calorie restriction makes the *SIR2* protein more active.

Guarente believes the answer lies in the ratio of NAD to a related molecule, NADH. Like NAD, NADH is found in species from yeast to humans, and it helps cells translate food into energy. Metabolic reactions in cells convert NAD to NADH, and vice versa. In 2000—soon after Sinclair joined Harvard's faculty—Guarente's lab reported in *Science* that with less NAD than normal, calorie-restricted yeast don't outlast regular yeast.

Then, in 2002, a *Nature* paper by Guarente's lab pulled NADH into the picture. It described how knocking out electron transport prevented calorie-restricted yeast cells from living longer. Electron transport is governed in part by NADH. "That really made it look like the NAD:NADH ratio would be a good candidate" for stimulating SIR2p activity, says Guarente, "but there was still no evidence one way or the other."

Sinclair's announcement at Cold Spring Harbor Laboratory that his data diverged from this theory took Guarente by surprise. In his talk, Sinclair presented an alternative model. His candidates were not NAD and NADH. Instead, he focused on a vitamin B precursor called nicotinamide, which is also a breakdown product of NAD, as well as a gene, *PNC1*, that converts nicotinamide into another molecule, nicotinic acid. Nicotinamide was already known to inhibit SIR proteins.

Sinclair found that without *PNC1*, calorie-restricted yeast didn't live longer than normal. Furthermore, adding copies of *PNC1* to yeast receiving normal amounts of glucose extended their life span. They behaved, in other words, as if they were on a low-calorie, low-glucose diet, even though they weren't.

Sinclair explains his model this way: *PNC1* senses when yeast cells are exposed to low glucose. That boosts the gene's expression, which depletes nicotinamide, which boosts activity of SIR2p, which extends life span. Mammals don't have a *PNC1* gene. But Sinclair believes that nicotinamide and genes that deplete it may guide SIR2p and related proteins in those organisms. *PNC1*, he notes, could also explain why other stressors such as heat shock extend life in yeast: The gene is upregulated by many environmental stresses, not just low glucose.

In May 2003 Sinclair's lab published this work in *Nature*. And to hammer home his point, he tested Guarente's theory and published a separate paper on the subject in *Science* last December. Its title didn't bode well for Guarente: "Yeast Life-Span Extension by Calorie Restriction Is Independent of NAD Fluctuation." The battle lines were drawn.

Guarente shot back. With his postdoc Su-Ju Lin, who recently moved to the University of California, Davis, he published a counterpoint in the January 2004 issue of *Genes and Development*. It addressed this question: Did the yeast strains Sinclair used, which lacked *PNC1* altogether and hence accumulated substantial levels of nicotinamide, mask fluctuations in the NAD:NADH ratio that occur in normal, although calorically restricted, yeast cells? In other words, Guarente wondered whether what Sinclair saw in his cells, while accurate, wasn't what happened in nongenetically altered yeast exposed to low glucose.

To test Sinclair's theory against their own, Guarente and Lin created yeast that lacked the *PNC1* gene. Everyone agreed that the cells' extra nicotinamide would inhibit SIR2p regardless of other factors, so they depleted the excess nicotinamide. When calorically restricted, the yeast still lived extra-long. The absence of *PNC1* didn't stop the cells from sensing calorie restriction and living longer as Sinclair's theory supposed.

Guarente did admit defeat in one arena. He agreed with Sinclair that NAD fluctuations weren't mediating *SIR2*. But he and Lin reported in their *Genes and Development* paper that the NAD:NADH ratio is crucial nonetheless. To their surprise, they say, calorie restriction appears to lower NADH levels rather than increase NAD. The drop in NADH, in turn, boosts the NAD:NADH ratio and extends life.

The response from outsiders to this burst of studies has mostly been bafflement. To begin with, no one can agree on whether Sinclair's and Guarente's theories are mutually exclusive, or whether they can coexist. (Even Guarente, Sinclair, and Lin don't agree on this score.)

"My gut view is that one can't be right," says Steven McKnight, a biochemist at the University of Texas Southwestern Medical Center in Dallas. A longtime fan of Guarente, his tendency is nonetheless to side with Sinclair—partly, he confesses, because the high ratio of NAD to NADH and the high NAD levels that Sinclair reports jibe with McKnight's own work.

The molecular biology that Guarente and Sinclair are tackling is so complex that biochemists have spent decades squabbling over some important details. Among them is a normal cell's ratio of NAD to NADH. McKnight falls into the school that endorses a high ratio of at least twenty, similar to what Sinclair reports; some others favor a much lower ratio of one to three, which Guarente stands behind. For NADH fluctuations to significantly impact the ratio, as Guarente postulates, the ratio must be low.

Richard Veech, a metabolism researcher at the National Institutes of Health in Bethesda, Maryland, takes issue with parts of both studies. "It's been well known since 1958 to any biochemist that [a ratio] of two or three is nonsense," he says of the Guarente paper. And as for Sinclair: "Our primary observations differ from his primary observations" when it comes to Sinclair's report that the ratio of free, unbound NAD to unbound NADH molecules has no impact on *SIR2* activity. (Guarente's paper measured the total levels of NAD and NADH, which includes molecules bound to structures in the cell.) Ultimately, though, Veech and some others conclude that more than one mechanism must be regulating *SIR2*. "You're going to control life span with one enzyme for one effect?" he says. "Please!"

BEYOND ADVANCING HIS OWN CASE in the *SIR2* clash, neither Guarente nor Sinclair is keen to discuss it. Both are now eyeing a world beyond yeast, pursuing mechanisms of aging in mammals. And both are chasing pet theories they hope will combat diseases of aging and potentially extend life.

In the year 2000, Guarente and his colleague Cynthia Kenyon, a worm researcher at the University of California, San Francisco, helped found Elixir Pharmaceuticals, which is a short walk from Guarente's lab in Cambridge. Roughly half the company's research is focused on *SIR2* and molecules that modulate its effects. (The other half revolves around a separate pathway identified by Kenyon in worms.) One of the toughest challenges in targeting the protein made by *SIR2*, however—or *SIRT1*, as it's known in mammals—is that "this protein is all over the body," says Peter DiStefano, the chief scientific officer of Elixir. The company is currently experimenting with various animal models and has raised more than $40 million from investors.

Elixir has also operated with almost no direct competition; pharmaceutical companies have hesitated to enter this market, and very few other biotechnology firms are devoted to aging research. If Sinclair has his way, that won't last long. In 2003 he asked Andrew Perlman, a twenty-eight-year-old millionaire who made his money selling two technology companies he founded, to help Sinclair build a new company called Sirtris Pharmaceuticals. Sirtris, which hasn't yet raised funds, will focus largely on Sinclair's most recent obsession—a compound called resveratrol, an antioxidant in red wine and other foods.

In a paper published in August 2003 in *Nature,* Sinclair and his colleagues reported that in yeast, resveratrol appeared to stimulate *SIR2,* hence mimicking calorie restriction and slowing aging. Various studies in animals also suggest that resveratrol protects against cancer. It's "as close to a miraculous molecule as you can find," says Sinclair. "One hundred years from now, people will maybe be taking these molecules on a daily basis to prevent heart disease, stroke, and cancer." A Montreal company, Royalmount, is beginning human trials of resveratrol in herpes and colon cancer prevention; Sinclair hopes Sirtris will partner with it. He's also experimenting with modified versions of the compound.

Because resveratrol occurs naturally, it's already widely advertised in health food stores and over the Internet. Sinclair purchased a dozen samples peddled as resveratrol and tested them in his lab. Only one passed the test—the compound is quite unstable at room temperature—and Sinclair briefly became a paid consultant to the company that makes it, Longevinex. In late December 2003 he announced that he had severed ties with Longevinex after the company broadcast comments from him on its Web site that Sinclair claimed were inaccurate.

Guarente, who tried to recruit Sinclair to Elixir before their falling-out, even bringing him to some of the company's board meetings, wasn't expecting to hear that his former postdoc was starting a company of his own. Sinclair's choices, however, mirror Guarente's years ago: As a young scientist, Guarente rejected an offer from his Harvard adviser, Mark Ptashne, to join Ptashne's new company. Instead, says Guarente, "I and a bunch of young turks at Harvard started a competitor company. . . . That's what young people do." The company eventually folded. Ptashne, he says, was "like my father, the establishment."

Certainly, Sinclair still views his mentor as something of a father figure. Asked if he's read Guarente's 2003 memoir, *Ageless Quest: One Scientist's Search for Genes That Prolong Youth*, he looks less than enthusiastic. "I don't want to see into his mind," says Sinclair. "It's a bit like learning about your parents' sex life."

Although many scientists agree that both the resveratrol molecule and the *SIRT1* drug target seem promising, they don't foresee smooth sailing. "I can't believe that it's going to be the magic bullet that cures aging," says James Joseph, a neuroscientist at Tufts University, of resveratrol; he has extensively studied the family of compounds to which it belongs. Resveratrol is also known to target a broad swath of molecules in the body, something that could be problematic in a drug, says Elixir's DiStefano. Sinclair hopes to avoid this by tinkering with the compound's chemical makeup.

The big question facing *SIRT1*, meanwhile, is what it does in mammals. The skepticism Guarente confronted in the early 1990s, when he backed yeast as a model for human aging, has abated, but fundamental mysteries remain. Perhaps most important, is calorie restriction in mammals mediated by the *SIRT1* gene? "It would surprise me only somewhat" if it's not, says Marciniak, a former Guarente lab member now at the University of Texas, San Antonio.

Both Guarente and Sinclair are wrestling with this question, and, as in other areas, they're racing along parallel paths. In early 2004 a paper in *Cell* by Guarente's team and a paper in *Science* on which Sinclair was an author both explored how *SIRT1* helps mammalian cells withstand environmental stress.

Guarente also talks animatedly about a project that's captured his attention: fat and its links to some of the seven *SIRT* genes in mice. He and his lab members are feverishly working to link these genes with fat accumulation and sensitivity to insulin—which could lead to new therapies for obesity and diabetes. "That's what I think the five-year plan is," says Guarente. He never imagined, he adds, that his work on life span might translate into diabetes drugs.

Sinclair is more coy, but he admits to exploring connections between fat and *SIR2* in worms and mice. The workaholic in him hasn't abated, even with a one-year-old daughter at home. He keeps a microscope, an incubator, and a refrigerator for yeast plates in his house. Reluctant to take time off, he's returned to Australia just once in the last three years. And Sinclair guards his work with extraordinary care: After a notebook filled with data went mysteriously missing, he installed a safe in his Harvard office. There the lab's notebooks sit, locked inside.

Guarente's lab is calmer these days than it was when Sinclair and others toiled there. "Now we have birthday cakes," says Marcia Haigis, a postdoc. "Lenny says the lab is too touchy-feely."

A more relaxed atmosphere hasn't stopped Guarente, like Sinclair, from sticking resolutely to his theory of aging. The truth, if and when it surfaces, may well embrace a synthesis of what the two—and others—propose. Or, of course, it may show that only one of them is right.

ROBIN MARANTZ HENIG

The Genome in Black and White (and Gray)

FROM THE *NEW YORK TIMES MAGAZINE*

Are there differences in race, beyond superficial features, that are rooted in our genes? As Robin Marantz Henig writes, it is a question that has major researchers divided, as new medications, designed especially to treat particular populations, return the issue, uncomfortably, to the forefront of scientific debate.

Imagine that you have heart failure. What can medicine do for you? It depends: are you white or black? If you're white, your doctor may prescribe one of the drugs that seem to ease the symptoms, maybe a beta-blocker or an ACE inhibitor. And if you're black, your doctor may still prescribe those drugs, but they might not really help.

That's about to change. In the not-too-distant future, if you're black and have heart failure, drug-company researchers predict you'll be able to go to the doctor and walk out with a prescription tailor-made for you. Well, not tailor-made, exactly, but something that seems to work in people a lot like you. Well, not a lot like you, exactly, except that they're black too. In this not-too-distant future, if you're black, your doctor will be able to prescribe BiDil, the first drug in America that's being niche-marketed to people of a particular race—our first ethnic medicine.

BiDil, expected to be approved in early 2005 by the Food and Drug Administration, is on the leading edge of the emerging field of race-based pharmacogenomics. It signals a shift in perception, a new approach to medicine that has at its core an idea at once familiar and incendiary: the assumption that there are biological differences among the races.

BiDil is also a feat of creative repackaging. In 1999 the FDA rejected it for use in the general population because it was found to be ineffective in the treatment of heart failure, a common complication of cardiovascular disease that affects some five million Americans and leads to three hundred thousand deaths a year. But in 2001 the manufacturer, NitroMed, asked permission to test BiDil exclusively in blacks, whose heart failure tends to be more severe and harder to treat. The company reasoned that the drug's effect on nitric-oxide deficiency, more common in black heart-failure patients than in nonblacks, might make it especially suited to them. With the collaboration of the Association of Black Cardiologists, NitroMed embarked on a large clinical trial involving more than four hundred black women and six hundred black men, all of whom had heart failure.

Last summer, investigators called an early end to the study because they thought BiDil was so effective that it would be unethical to continue to deny it to people in the control group. Thus, a drug that had been deemed ineffective in the population at large seemed to work so well in one racial subgroup that the scientists thought everyone in that subgroup should get it.

Pharmacogenomics has for years been touted as the ultimate benefit of the genomics revolution. But to many, this revolution has a troubling side. For race-based niche marketing to work, drug developers first will have to explore the ways that blacks, whites, Asians, and Native Americans are biologically different. And the more they explore and describe such differences, critics say, the more they play into the hands of racists. Even the broad-minded might inadvertently use such information to stigmatize, isolate or categorize the races. Could it be that this terrain is too dangerous to let anyone, no matter how well meaning, try to navigate it?

In the fall of 2004 a major scientific journal, *Nature Genetics*, published a special issue on the genetics of race. This came on the heels of several conferences on the subject, as well as editorials in the science press, including one in the *Journal of the American Medical Association* that appeared just weeks before the *Nature Genetics* special issue. All of these forums pose some thorny questions: Can genes tell us anything meaningful about race, beyond the obvious connection to things like skin color? Do the races differ biologically in terms of drug response or disease susceptibility? Can genes say anything about how "race"—which is itself all but impossible to define—is related to complex traits like behavior and intelligence?

Looking for biological determinants of race is nothing new. It has a potent history, with poisonous associations dating back to the early days of eugenics. But contemporary science has given these efforts a new respectability. In the wake of the completion of the Human Genome Project, geneticists are trying to arrange pieces of the genome like a Rubik's Cube, searching for patterns of variation that align into some useful matrix. Their goal is to generate information that will help prevent and treat common diseases. But in the process, they're generating information that might also lead to declarations about the biological meaning of race.

The new interest in racial genetics comes at a time when the softer sciences, like anthropology and sociology, have declared that race is a cultural construct, without any biological significance. The social designations go back at least to the nineteenth century, when humans were generally divided into five races that were loosely tied to skin color; this has lingered as the basic grammar of race even into the twenty-first century. But in a 1998 position paper, the American Anthropological Association called race a social invention, with a variety of pernicious consequences ranging from day-to-day bigotry to the Holocaust. Racial beliefs are myths, the anthropologists wrote, and the myths fuse "behavior and physical features together in the public mind, impeding our comprehension of both biological variations and cultural behavior, implying that both are genetically determined."

Geneticists, too, have gone on record as saying that race has no biological significance. "The concept of race has no genetic or scientific basis," said J. Craig Venter in June 2000, standing beside President Bill Clinton to announce the completion of the first draft of the human genome sequence. Venter was at the time the president of Celera, the private company that competed with the National Human Genome Research Institute, a publicly financed international team, to sequence the genome. (It was declared a tie.)

Venter's scientific rival, Francis S. Collins, the head of the genome institute, stood at the podium that day on Clinton's other side—two male, middle-aged white scientists saying we're all brothers and sisters under the skin. Collins made much of the fact that humans share 99.9 percent of their genome with one another—and that the remaining 0.1 percent probably codes for variations, like skin color, that are for the most part biologically insignificant. In fact, there is more variation within races than between them. A few months later he made the point more informally, playing his electric guitar and regaling his coworkers with a musical ditty he had written to the tune of Woody Guthrie's "This Land Is Your Land":

> We only do this once, it's our inheritance,
> Joined by this common thread—black, yellow, white or red,
> It is our family bond, and now its day has dawned.
> This draft was made for you and me.

Today, the two men have parted company on this narrow strip of common ground. Venter says he still believes the genome is colorblind. "I don't see that there's any fundamental need to classify people by race," he says. "What's the goal of that, other than discrimination?"

But Collins sees the matter differently now. Maybe in that 0.1 percent of the genome there are some variations with relevance to medicine, he says. And maybe identifying them could help reduce health disparities among the races. He is using his bully pulpit at the genome institute to urge scientists to study whether these variations can, or should, be categorized according to racial groupings.

"It's always better to face up to a controversial scientific issue, to tackle the issue head-on and not run away from it," Collins says. "And if we don't do it, someone else will—and probably not as well."

ONE REASON TO FOCUS on the genetics of race is to try to make a dent in health disparities: the frustrating gap in the health status of different racial groups that stubbornly refuses to close or even to be adequately explained. In terms of national measures of physical well-being—life expectancy, infant mortality, some chronic diseases—blacks tend to do worse than whites. Many factors account for this health gap, including the fact that minorities suffer disproportionately the effects of low income, lack of health insurance, poor diet, exposure to environmental toxins, discrimination, and stress. But some geneticists think that at least some part of health disparities can be explained by genes. Social scientists think genetic explanations might obscure the all-too-real social and economic causes.

Take hypertension, which affects black Americans at a higher rate than white Americans. Geneticists try to explain this difference in terms of genes: genes for salt retention, genes for low levels of renin in the kidneys. But a classic study found that one thing that correlated most strongly with level of blood pressure was, surprisingly, skin color. Among black subjects of low socioeconomic status, the darker the skin, the higher the blood pressure. Social scientists' explanation is that people with darker skin are subject to greater discrimination, and therefore to greater stress.

"If you follow me around Nordstrom's, and put me in jail at nine times the rate of whites, and refuse to give me a bank loan, I might get hypertensive," says Troy Duster, a professor of sociology at New York University and at the University of California, Berkeley. "What's generating my increased blood pressure are the social forces at play, not my DNA."

But pharmacogenomics researchers presume that health disparities can be addressed, at least in part, by exploiting tiny group differences in DNA. If the BiDil experience pans out, other companies are likely to try their own versions of race-based drug development. Some candidates already exist. People known as slow acetylators, for instance, take a longer time than fast acetylators to clear certain drugs from the liver. This means they're more likely to build up toxic levels of some common drugs. The proportion of slow acetylators in different racial groups ranges from a low of 14 percent among East Asians to a high of 54 percent among whites. Some whites, therefore, might benefit from a different version of medications that are cleared through the liver.

The ultimate goal of pharmacogenomics would be for everyone's genome to be analyzed individually, so that doctors could gauge how much of a medication, and which type, is most likely to work for a specific patient. Even the BiDil investigators are moving in that direction. Michael D. Loberg, the president of NitroMed, says that the company asked each participant in the BiDil trial for permission to take a DNA sample and that he hopes to get a total of at least four hundred such samples. These will be sequenced, he says, "to see if there's some genetic marker that predicts which of the trial patients responded to BiDil favorably and which didn't."

But at this point, geneticists cannot sequence individual genomes in a cost-effective way. Until they can, they may view race as a handy shortcut, a way to make some useful generalizations about how an individual patient will fare with a particular drug. But while using race this way might increase the odds of finding the right medication, it is an imprecise method, a kind of roulette in which the physician is making educated guesses based on probabilities.

The temptation of race-based medication is clear: it's convenient for the investigator, and it suits the way drug companies' products are sold. "The mantra of pharmacogenomics is that drugs will be fine-tuned for the individual," Duster says. "But individuals are not a market. Groups are a market." And one typical way to identify markets, in a country where skin color seems to count for so much, is race.

In terms of our genes, we humans are all the same—except for the ways in which we're different. The human genome comprises three billion nucleotides, strung together in a specific order along the chromosomes. About 99.9 percent are identical from one person to another, no matter what that person's race, ethnicity, continent of origin or bank account.

Among our three billion nucleotides, an estimated ten million are locations of common variations. Where most people will have a nucleotide represented by the letter A, for instance, a big group of people might have a T instead. Elucidating where those spots are, and whether replacing a T with an A has any clinical significance, are what occupies today's geneticists.

The most common type of variants are called single nucleotide polymorphisms, or SNPs (pronounced "snips"). Usually they occur in regions where the nucleotides seem to be doing nothing. This means the SNPs don't have any function, either, or at least none that has been discovered yet; they're just there.

Still, SNPs tend to occur in different patterns in different populations. Say there's a SNP on Chromosome 12 in which a person might have either an A or a T. At this hypothetical SNP, 20 percent of Africans might have an A, and 80 percent a T. At the same spot, the frequency might be flipped in Europeans: 80 percent might have an A, while only 20 percent have a T.

So while SNP patterns don't reveal anything about the function of the genes, they can say something about an individual's continent of ancestry—and, by extension, something about migration pathways through human history. SNPs tend to be inherited in clusters, called haplotype blocks. Like SNPs, varieties of haplotype blocks occur at different frequencies in different regions of the world—and that's how population geneticists have managed to reconstruct the story of human migration.

The biggest variety of haplotype blocks occurs in Africa, because modern humans arose there more than 150,000 years ago, and variations have had the longest chance to accrue simply because of random mutations. About 55,000 years ago, a small group of modern humans, who carried in their genomes a subset of the original haplotype varieties, traveled to Australia; later, in sequence and timing that are still a source of controversy among paleoanthropologists, other small groups migrated to parts of Europe, Asia, and the Middle East.

As time went on, there were some evolutionary changes in response to the new environments. In Northern Europe, for instance, people carrying mutations for lighter skin color thrived, probably because the scarcity of sunlight made dark-skinned people especially susceptible to Vitamin D deficiency and rickets. But most of the variations occurred in the nonfunctional regions of the genome, with no effect on an individual's appearance or health. All that the variations did was allow geneticists, some two thousand generations later, to assign a continent of origin to the descendants of these original travelers based on the descendants' DNA.

TO THE DISMAY of Troy Duster, several private companies are now taking these findings about SNPs to a new level: scanning the genome for variations that can say something about an individual's race. Last year, a company called DNAPrint Genomics made headlines by telling law-enforcement officials in Louisiana that they'd been looking for a serial killer of the wrong race. Eyewitnesses had offered different accounts of the race of the suspect—some thought he was black, others white—and authorities had focused their search on white males between the ages of twenty-five and thirty-five based in part on an FBI psychological profile. But based on crime-scene specimens, DNAPrint said the murderer was probably black—in fact, the company said it could detect 85 percent sub-Saharan African ancestry and 15 percent Native American—and even gave an assessment of his skin tone. When a black male was apprehended, his DNA was found to match that at one of the crime scenes. He was convicted of second-degree murder in August.

For some, this would be a story of science advancing police work. But for people like Duster, the forensic use of genetic markers raises troubling questions. Can a DNA screen of a person's blood or hair really tell you anything more than where his ancestors probably came from? Would it lead to witch hunts based on some uncertain appraisal of skin color? Would it be used, wrongly, to give a patina of scientific authority to group prejudices?

Worried, Duster approached his friend and colleague, Francis Collins, to suggest that Collins might want to use his position at the genome institute to mount an investigation into the genetics of race—before the drug manufacturers and genomics companies set the tone for the public debate.

Collins says he was already thinking the same thing. The two men approach the venture from different perspectives, less because Collins is white and Duster is black than because one is a geneticist and the other a sociologist. As Duster sees it, race is a relationship, largely dependent on social context. Take a Tutsi and a Hutu and set them down in Los Angeles, he says, and they're both the same race, both black. But put them back in Rwanda, and they're two different races, different enough to slaughter each other.

There may be biological dimensions to race, Duster says, but that doesn't take away from his belief that race should be understood as a social construction. "The myth is that somehow the biology is real and the social forces are unreal," he says. "In fact, the social forces can feed the biological forces."

Collins, for his part, recognizes that social forces explain many of the observed differences among the races—but says he thinks something else might be involved as well. "We need to try to understand what there is about genetic variation that is associated with disease risk," he says, "and how that correlates, in some very imperfect way, with self-identified race, and how we can use that correlation to reduce the risk of people getting sick."

Taking up Duster's challenge, Collins knew, meant walking into a quagmire. A decade earlier, another top government scientist lost his job by discussing the genetics of urban violence (though his case was egregious: he compared young black men with male monkeys). But Collins said he believed the idea, risky as it was, was worth pursuing because it offered the best chance of converting new genomic information into something of medical significance.

The genome institute, part of the National Institutes of Health in Bethesda, Maryland, currently spends some $31 million for studies into human genetic variation. The institute is also a major contributor to the Hap Map project, a $110 million international collaboration that by late 2005 will have put together a coherent almanac of human variation using haplotype blocks. The Hap Map is meant to help scientists in their search for common disease-causing genes, but in the process it will also generate new information about the specific ways in which populations from the places being studied—China, Japan, Nigeria, and, in the United States, Utah—differ from one another genetically.

Collins is clean-cut and homespun, emphatically tall, with a fringe of sandy hair that makes him look younger than his fifty-four years. He exudes an aw-shucks earnestness when he talks about his favorite topics, which include his rebirth as a Christian during his medical training. Each time he makes a scientific discovery, he says, he gets a glimmer of insight into the workings of the mind of God.

But for all his personal sincerity, Collins is finding that some of his allies are wary of this newest undertaking. They know that even a man with the best intentions can muck it up when it comes to race.

WHILE WRITING THIS ARTICLE, I took a trip to the Holocaust Memorial Museum in Washington. I wanted to see the museum's exhibit about eugenics, the scientific movement of the early twentieth century that looked for evidence of biological racial differences to promote creation of a "fitter" species. In a very short time, eugenic ideas were subverted to support Nazi policies of ethnic cleansing and racial extermination. Since the spring of 2004, when Collins called to suggest that I might be interested in his institute's plan to investigate the genetics of race, I had talked to more than two dozen scientists about the issue. Uncomfortable questions about where such inquiries could lead underscored a number of those conversations—the sort of questions that, as a white person in America today, I don't usually have to confront. I went to the Holocaust museum looking for resonances.

How disturbing it was to see that the activities of the early eugenicists resembled, from a certain perspective, the activities of specimen collectors of the early days of zoology—as well as those of genomics researchers today, going around collecting specimens of human variation. The eugenicists engaged in some straightforward scientific studies that can seem almost harmless, even ordinary. And that's what makes it so troubling to look back.

With rulers, calipers, charts of eye shapes and elaborate reconstructions of family trees, eugenicists of the 1920s and 1930s took great care to describe physical characteristics of different racial groups. They photographed subjects, measured their noses and mouths, made plaster casts of their faces and documented variations in facial features and head proportions. Is it possible that the difference between then and now is that the tools have changed—that instead of using calipers and scales, scientists now use DNA-sequencing machines?

Connecting contemporary genomic studies to the Holocaust is too glib, of course, and it obscures one crucial point: that the anthropometrics of the early eugenics movement turned ugly once fanatics perverted the information. But the exhibit is a sobering reminder of how easy it would be to travel down that path. "I think our best protection against that—because this work is going to be done by somebody—is to have it done by the best and brightest and hopefully most well attuned to the risk of abuse," Collins says. "That's why I think this has to be a mainstream activity of genomics, and not something we avoid and then watch burst out somewhere from some sort of goofy fringe."

Collins doesn't quote the Bible often—he tends to neither hide nor flaunt his religious faith—but he quotes it now. He chooses a line from the New Testament's Book of John, in which Jesus says to his disciples, "And you shall know the truth, and the truth shall set you free."

Reducing health disparities and catching criminals are serious reasons for pursuing the genetics of race, but there's also a small but growing trend toward something its practitioners call "recreational genomics." To satisfy curiosity about their heritage, more and more people are experiencing race-based genomics as a mail-in test, for which they pay up to four hundred dollars, that will tell them how much of their genome is black, white, American Indian, or Asian. These companies go beyond old-fashioned genealogical services, the kind that involve scouring archives and huddling over microfiche machines, and trace genetic linkages back many generations to a particular geographic location. Critics say that what these companies are doing sanctifies the genetic distinctions among racial groups, as if the question of whether race has a biological basis has been settled.

The services, with names like GeoGene, AncestryByDNA, and Roots for Real, begin by asking clients to mail in a cheek swab to get some stray skin cells from which DNA can be extracted. Though the process may feel like a parlor game, the results can be deeply affecting. One of those who traced his genetic lineage through a company called African Ancestry is Andrew Young, former United States ambassador to the United Nations and now chairman of an organization called Good Works International.

Young was looking for information about his maternal lineage only; he assumed, he told me, that his paternal lineage would be "contaminated" with white DNA, a bitter memento of slave rape that he didn't feel ready to confront. (According to Rick Kittles, a cofounder of African Ancestry and a geneticist at Ohio State University, about one-third of blacks who do a paternal lineage analysis, himself included, find that there is European DNA somewhere in their past.)

When a black client discovers that there's white in his genome, the results can be shattering. In 2003 the ABC News program *Nightline* profiled a fifty-year-old California man who had assumed his whole life that he was black. But a recreational genomics analysis by DNAPrint Genomics indicated that his genome was 57 percent of what the company called Indo-European, 39 percent Native American, 4 percent East Asian—and zero percent African. So what is this man: the race he has always thought himself to be, or the race his genome says he is?

Young's reaction to his African Ancestry report is an indication of how much weight we ascribe to genes, how much we believe our DNA reflects not only our racial identity but our individual identity too. When Young heard that the company had traced his DNA back to what is now Sierra Leone, he was disappointed because he considered Sierra Leone to have a "snobbish" middle class. But the report got much more specific: the people whose SNP pattern most resembled Young's, it said, were from the Mende tribe. Whether a few SNP matches can allow such precision is a matter of debate, but it fit happily into Young's self-image. Young, who got his start in the civil rights movement, was raised on tales about the *Amistad* slave-ship rebellion of 1839, for which the Mende were responsible. "I always had a spiritual connection to these stories," he says. "Now I have a genetic connection."

So is there such a thing as race? It depends on whether you're defining it in terms of culture or biology. Culturally, there is no denying it. In the United States, with its race-stained legacy dating back to slavery, the government has tried for centuries to define a person's race. The Census Bureau has been asking about race on its forms since the very first census in 1790, most recently giving individuals the opportunity to check off more than one race if they so desire.

But the more vexing question is whether there's such a thing as race in terms of biology. Genetic variations do seem to cluster differently for people with different continents of origin, but is this race? And what does it mean if it is—or if it isn't? Do we need to agree on whether race is a biological entity, since we can so readily agree that it's a social one?

"Race is a reality in this country, no matter what the genome tells us," says Vanessa Northington Gamble, director of the National Center for Bioethics in Research and Health Care at Tuskegee University. "If I can't get a cab in New York, it's because my skin is black. And I can't hold up my DNA and say: 'Wait a minute! I'm just the same as you!' "

Some critics worry that the more we find out about genetic differences among people of different racial groups, the more such information will be misinterpreted or abused. Already there are fears that the biological measures of racial differences might lead to pronouncements about inherent differences in such complex traits as intelligence, athletic ability, aggressiveness or susceptibility to addiction. Once such measures are given the imprimatur of science, especially genomic science, loathsome racist stereotypes can take on the sheen of received wisdom.

Looking for racial genetic markers does indeed risk creating categories that can get us in trouble. It bears remembering, however, that the "slippery slope" argument is itself a danger. Rather than abort a whole field of research because it might bolster cranks and demagogues, maybe one solution to our national angst over race is to let scientists hunt down the facts—facts that will no doubt affirm, one way or another, that the human genome is indeed our common thread.

Gods and Monsters

FROM *MOTHER JONES*

While so-called genetic engineering may bring cures to intractable diseases, it could also create strange, even frightening, hybrid creatures. Mark Dowie finds a biologist who is fighting against these "chimeras"—by trying to patent them.

On April Fools' Day 1998, within hours of reading U.S. patent application No. 08/993,564, the Honorable Bruce Lehman did something no other commissioner of patents had done in the two-hundred-year history of America's oldest government agency. He stepped before a cluster of microphones and announced that the patent would never be approved. No half-human "monsters" would be patented, Lehman declared angrily, or any other "immoral inventions."

Legal scholars—accustomed to an office bound by statute to remain silent until patents are approved or rejected—were shocked. Forgoing the traditional eighteen-month review period, Lehman had issued a marching order to his staff to reject a patent application they had barely read, rather as if a judge had instructed a jury that the defendant was guilty before the trial began. Furthermore, to support his decision, Lehman cited an 1817 court ruling that excluded inventions "injurious to the well-being, good policy, or good morals of society." But patent law had long since been amended to say that if an applicant could claim constructive use for a patent, he or she could not be denied simply because there might be dangerous or unethical uses of the invention.

"Even attorneys who worshipped the system were horrified," recalls former patent examiner Peter di Mauro, who has since left the agency. Research biologists and biotech executives also felt blindsided, hearing in Commissioner Lehman's outburst a threat to the hard-earned clearance they had won from the Supreme Court eighteen years earlier to patent "anything under the sun made by man"— even living organisms.

Strange as it may seem, the inventor, Dr. Stuart Newman, a soft-spoken developmental biologist and professor at New York Medical College in Valhalla, New York, completely agreed with Lehman that his invention defied the boundaries of human morality. It's why he filed for the patent. And it's why, six years later, as the biomedical community holds its breath, he and the Patent Office remain locked in a legal battle that may redefine what we mean by "human."

NEWMAN'S PATENT APPLICATION is for an intriguing biotechnological contrivance called a chimera [**ki**-mir-a]. According to Greek mythology, a chimera was a part-lion, part-goat, part-serpent creature that terrorized Lycia until it was slain by the hero Bellerophon. If biotech continues to run amok, warns Newman, such inventions of legend and allegory could actually *be* invented.

Created by injecting the embryonic stem cells of one or more species into the embryo of another species and then allowing that embryo to continue development in the womb of either species, a biological chimera is a way to hybridize two or more species that won't cross sexually. The exact results are largely unpredictable except for the certainty that the chimera will contain cells of each species proportionate to the numbers placed in the embryo. A creature made from an equal number of cells from two species could look like one species but contain the genes, organs, and intelligence of the other.

Newman seeks to patent "chimeric embryos and animals containing human cells." And while his application cites innocuous biomedical uses for human/animal chimeras—such as toxicological research and the potential for growing rejection-proof human organs in pigs or other creatures—taken to its most extreme but not necessarily impossible end, the technology could be used to manufacture soldiers with armadillolike shielding, quasi-human astronauts engineered for long-range space travel, and altered primates with enough cognitive ability to ride a bus, follow basic instructions, pick crops in 119 degrees, or descend into a mine shaft without worrying their silly little heads about inalienable human rights and the resulting laws and customs that demand safe working conditions.

At first blush, what Newman seeks sounds quite like many patents already obtained by university laboratories and biotech corporations to insert human genes into mice and other mammals, creating what is known as a "transgenic" animal. But cross-pollinating using whole cells containing the entire genomic sequence is a profoundly different and even more morally charged process, and Newman's invention presents the Patent Office with a serious legal and political quandary that could earn the agency enemies either way it might rule. Granting a patent for a half-human chimera would throw religious, bioethical, animal-rights, and constitutional activists into high dudgeon. And the biotech industry would boil over the approval of what is clearly a preventive patent. But decline it and the agency is in court, eventually the highest court in the land. The last time that happened the Patent Office lost its case. Thus far, however, Commissioner Lehman and his successors have decided that foiling Newman is worth the risk.

And Newman, a man accustomed to disappointing reversals in and outside his laboratory, has fought back, claim by claim, for six years, because he knows that with a patent in hand he can delay what he regards as a deeply offensive technology for the twenty-year life of the patent. Opposed to genetically altering human beings and patenting living organisms, Newman (who's supported in this endeavor by techno-gadfly Jeremy Rifkin) sees in the chimera the manifestation of all he finds immoral about biotech. The technology exists to make chimeric embryos "tomorrow," he says, and a chimera of two similar species (say human and chimp) that could survive into adulthood probably isn't far off. "This is the prize. The more you can humanize animals genetically, the better they are for research models and as sources of transplantable tissues."

With such goals in mind, six Canadian and American biologists gathered November 13, 2002, at the New York Academy of Sciences to debate in private whether or not to proceed with chimeric research involving human cells. While two of the scientists raised questions about the ethics of such research, the rest felt that while it could admittedly lead to outcomes "too horrible to contemplate," chimeras still offered enough medical promise to proceed. Dr. Newman has since notified all six that if they proceed in developing chimeras, they will be in violation of his pending patent, and litigation will follow.

News of the meeting, immediately reported in *Nature*, threw the Patent Office into an even deeper bind. "This goes way beyond the jurisdiction of the agency," Deputy Commissioner Stephen Kunin said. "We're being dragged into a controversy which, from our perspective, we don't need to be part of."

In truth, the Patent Office made itself part of this controversy almost two decades ago when it began granting patents for genetically altered mammals. At that moment, it placed itself and the patent process at the intersection of science, commerce, and religion, and in Stuart Newman's line of fire.

THE PATENT ACT OF 1793, drafted by Thomas Jefferson, and subsequent case law stipulated that nature, in any of its forms or manifestations, could not be patented. But in the late 1920s research agronomists approached Congress and argued that man-made plant hybrids were not really products of nature, but rather existed as a consequence of human manipulation and were thus, in Jefferson's carefully chosen words, "a new composition of matter." The 1930 Plant Patent Act soon allowed for new varieties of plants that reproduced asexually to be patented.

Molecular biology soon produced other techniques for creating new organisms and genetically altering existing ones. But most required sexual reproduction and were thus deemed "natural." Then on June 16, 1980, the U.S. Supreme Court surprised both sides in an obscure patent case by ruling five to four that a General Electric scientist named Ananda Chakrabarty could patent genetically modified bacteria that produced enzymes capable of breaking down crude oil. In its opinion, the majority concluded that Chakrabarty's bugs, with their vibrant cells and shimmering DNA, were "a human-made invention . . . with markedly different characteristics from any found in nature." They were new compositions of matter, and were, for the life of the patent, the sole property of General Electric, which, fearful of releasing such an engineered life-form into the environment, never used them.

Soon, thousands of life-form patent applications were before the Patent Office, which rapidly evolved from an agency opposed to patenting life-forms to an outright booster of the practice. Biotech scientists, executives, and their attorneys surmised that if whole organisms were now patentable, then surely inert pieces of them—genes and gene sequences—would be as well.

A 1986 patent for corn genetically manipulated to produce tryptophan—the chemical the human body uses to make serotonin—widened the opening. In 1987 the Patent Board of Appeals rejected a process to produce bigger oysters through pressure as "too obvious" but noted that involvement of a multicellular animal was not itself a bar to patentability. On the strength of that decision, Patent Commissioner Donald Quigg called a press conference and announced that all "multicellular living organisms, including animals," were patentable. He did make a specific exception for human beings, a restriction mandated, Quigg said, by the Thirteenth Amendment, which prohibits their ownership.

Few realized how much wider the Quigg decision had opened the pathway to life-form patenting until a year later, when after a five-year court battle, the Patent Office issued Patent No. 4,736,866 to Harvard biologists Philip Leder and Timothy Stewart for mice transgenically engineered to develop tumors. The famous OncoMouse was born, and DuPont, which funded the research, holds the exclusive license to it and all its progeny, though the National Institutes of Health is using the mice for cancer research with the understanding that any commercial application it develops belongs to DuPont.

The OncoMouse was the first complete, living, breathing mammal to be patented in the United States, but it was not the last. Scores of transgenic mammalian patents have since been approved, including those for cows and sheep modified to produce medicinal milk ("factory pharming"), rats with a propensity to develop inflammatory diseases, and a mouse whose cells contain the gene for human insulin. When Newman filed in 1997, his was among fifteen-thousand biotech patents applied for that year. By 2002 the number had nearly doubled. Most are approved, many of them conflicting with existing patents, leaving previous owners no recourse but the courts, which thrills patent attorneys. And, despite Quigg's Thirteenth Amendment stipulation, applications to patent human genes also increase, even genes for which there is no known purpose or product. And they are approved. When objections are raised to this practice, particularly by scientists, the Biotech Industry Organization argues that the United States would lose its "global competitive advantage" if a patent slipped away. "Few people realize," one newspaper article notes, "that a genetic 'land grab' is taking place on a scale that rivals nineteenth-century colonialism."

IF NEWMAN'S PATENT seems the stuff of fancy or *Planet of the Apes,* consider that a sheep/goat chimera was made back in 1984. It's called a "geep." When fully grown, it looks a little like a sheep with features of a goat, or vice versa, and the cells of each are fairly evenly scattered throughout the geep. In a chimera, the two sets of cells remain distinct (unlike in a hybrid such as a mule, where all the cells are hybridized "mule cells"), ensuring that a geep will accept skin grafts from either the sheep or the goat whose cells it owns without stimulating an immune response. And it can be used to test the toxic effects of chemicals on both sheep and goats. Thus a part-human chimera, even one without obvious human characteristics or rights, could be used to test for human carcinogens and mutagens without the safeguards currently required for clinical trials.

In his application, Newman says "the cells composing the embryo may contain one or more transgenes." With such a refinement, chimeric technology could one day be used to engineer specific tissue characteristics of one species into the organs of another. That would be the first step toward growing transplantable human organs inside other animals, a product of enormous commercial interest to biotech companies like Advanced Cell Technology (ACT), which looks on with envy and distress as Newman does battle with the government.

A Boston-area company, ACT is already inserting human DNA into cow embryos to produce and patent human stem cells. It is also attempting to design and breed pigs that will look, smell, and behave like pigs in a feedlot but possess hearts, livers, kidneys, and pancreases so humanlike in size, shape, function, and tissue type that people will not reject them after transplantation.

Either way the Patent Office rules on Newman's application is bad news for ACT. Approval means Newman owns the technology they need to advance their chimeric research, and ACT executives know that such an activist is unlikely to license it to them or anyone else. Rejection by the Patent Office sends a chilling signal through the entire biotech industry that the government has found a moral boundary beyond which it is unwilling to tread.

This may explain why ACT's high-profile vice president of scientific and medical development, Robert Lanza, told me that he supports Newman's effort "100 percent" but added, "While I am opposed to creating chimeric animals, I do not approve of tying up any technology that could save lives."

ALTHOUGH IT IS NOW THE EXPLICIT GOAL of the Patent Office to complete patent reviews within twenty-four months, Newman's patent is still officially pending, six years after it was filed. On four occasions patent inspectors have rejected his application, but with objections so vague and specious that Newman and his attorneys have been able to argue it back to life, only to be rejected again on new grounds. When the agency objected to using human embryonic cells to help create a chimera, for example, Newman's lawyers pointed out it is legal to abort 100-percent-human embryos, and it would make no sense to grant part-human embryos greater protection.

But one objection survives each exchange—that the chimera itself would be too humanesque. All the recipes in Dr. Newman's application call for using human embryonic cells and stem cells, and describe methods likely to raise the ire of anyone who believes that human life begins at conception or that human cells are more sacred than those of other creatures. Newman's chimera could end up looking, feeling, and behaving somewhat like a human being. That so troubled patent inspector Deborah Crouch, the first of four examiners to work the case, that she coined an intriguing expression to justify rejection. "Since applicant's claimed invention embraces a human being," she wrote, "it is not considered to be patentable subject matter." Newman and the Patent Office have spent much of the subsequent six years grappling with the meanings of "embrace" and "human."

To make matters more bizarre, while Newman and his lawyers argue against the embraces-a-human assessment, Newman in fact agrees with it. He knows at least four ways to make a chimera, but he has no intention of ever making one, with or without human cells and features. So as not to disseminate technology he considers dangerous, his application includes only techniques to make chimeras that can be found in existing literature, which gave examiners another excuse to reject his patent—lack of originality. Newman responded that it is not the method for which he seeks a patent but the "unique application" of using human cells and, of course, the final product.

What that product would be, exactly, is at least as much the stuff of philosophy as science. Place a human gene or gene sequence in a pig or dog and you haven't really moved either animal very close to being human as we understand it. But place the same genes in a chimpanzee, whose DNA is so similar to a human's that only a full genomic scan can tell them apart, and you could create something almost human, something that perhaps begins to resemble or "embrace" a human. This would be even more likely if you made a new creature by combining embryonic cells, rather than just the genes, from two or more species, as in Newman's chimera. "Different people would draw the line between nonhuman and human at different places," says Newman. "The problem is, the material continuity among all living organisms is such that no matter where you choose to draw it, based on whatever philosophy and belief system, the technology will eventually enable crossing that line."

Pull back from the molecular level and the answer to the question "What is human?" is fairly obvious. Bruce Lehman echoed Justice Potter Stewart's famous definition of obscenity when asked what he thought would constitute a chimeric human. "I'm quite certain that when we see one of these, we'll know it," he told the *Washington Post* a few weeks after issuing his summary rejection in 1998. But with a chimera's ability to hide distinct human features, even intelligence, inside another species, the definition of humanity could be blurred beyond recognition.

Now that he's stepped down from the Patent Office, Bruce Lehman can talk about pending patents, and so in August 2002 we met a few blocks from the U.S. Capitol in his modest office at the International Intellectual Property Institute, an organization he created after leaving office with a $1 million grant from the Bush administration. IIPI promotes the installation of patenting systems in every country of the world. As a private citizen, Lehman is free to reveal his true motivations for stifling Newman's patent, and as a patent evangelist he's happy to expound on the vital role intellectual-property laws play in industrial development, economic globalization, and the commercialization of biotech inventions. And he is more than willing to rebuke bio-Luddites such as Newman, whom he regards as anti-science.

"Stuart Newman is promoting an effort that will make it difficult to engage in biological research and commercialize the fruits of that research," says Lehman. "It's not funny or cute; it is profoundly wrong. Every attempt to stop science has been characterized by darkness."

Lehman says he was acting on the concept of *ordre public,* a European legal restriction against immoral inventions that does not exist in American patent law. Nevertheless, he believes that every American public official should be responsible for defending ordre public whether or not it is required by statute.

Yet Lehman also told me that had Advanced Cell Technology, Geron Corp., or almost any biotech firm applied for the same patent Newman did, he would not have stood in their way. And though he spoke of "monsters" when he first reacted to Newman's patent, Lehman now says he does "not believe there should be a prohibition against a human patent." It was not the specter of half-human chimeras or even patenting whole human beings that revolted him, he now says. "I was just deeply offended by anyone attempting to use the U.S. Patent Office to make a point, or to stop the advancement of science. I refused to make it easy for him."

Whatever his motives, by violating the agency's rules and issuing a premature and emotional ruling, Lehman may have given Newman adequate justification to appeal a rejection. By law, inventors are entitled to revise their applications and reapply as often as they wish in pursuit of approval. But Newman and his lawyers have decided that if the agency persists in rejecting the invention, they will appeal the decision all the way to the Supreme Court.

NEWMAN AND HIS ATTORNEYS have had only one audience with patent examiners. The "interview" occurred in January 2001, when the case was transferred to a new examiner. "They really didn't know what to make of us," recalls his attorney at the time, Patrick Coyne, who says they talked through the "embracing" issue and tried to explain to the new examiner why the invention was "original."

For the moment, patent officials seem content to leave Newman's application in limbo while they strategize their next move. They're faced with four options: They can withdraw all their objections and grant a patent; they can withdraw legal and scientific objections and reject the invention because it "embraces a human"; they could drop the embraces-a-human complaint and reject the patent for lack of originality or utility; or they could stand their ground on all objections and let the courts decide.

There are vexing questions at work here, both moral and legal. Should new organisms be created? Should any organisms be patented? And should humans or anything remotely resembling humans be created or patented? While it is really only the business of the Patent Office to deal with the second question, Newman's initiative could eventually force the entire federal government—executive, legislative, and judicial branches—to consider the others.

In their third response to Newman, sent in August 2000, patent officials acknowledged, "In the absence of clear legislative intent and guidance from the courts, it is incumbent on the Office to proceed cautiously." That is tantamount, contends attorney Coyne, to admitting that the agency has been granting life-form patents for more than twenty years without congressional authority. Newman hopes this means that Congress and the courts may soon be forced to reconsider the entire patent law, which was last rewritten in the early 1950s, long before the biotech era began.

"They're asking themselves how far they can go," Newman says. "They seem set upon turning down the application as an act of prudence and letting me take it to the Board of Patent Appeals and beyond." Patent Appeals is an interagency court where inventors and patent examiners can haggle before administrative judges over what is or is not feasible, novel, and useful about an invention. But in this case, it is also a place to decide whether to grant a handful of biotech companies control over the genetic blueprint of evolution and open the larger debate over what is and is not human. For while Congress and the courts have made clear that humans are not patentable, they've never defined the term "human."

It's hard to read any legal mind, especially on the current Supreme Court, but if the court does eventually hear the case, some arguments and counterarguments can be anticipated. One justice might argue, for example, that chimeric mice possessing human brain cells are still mice, much as the Patent Office decided that the OncoMouse, with its array of human genes, was a mouse—period. Bring a cage full of the brainy little squeakers before any court and the point will be obvious. They're rodents.

However, another justice would surely observe that if human cells become scattered throughout the entire body of a gorilla, a primate that is genetically 98 percent human, and the chimera looks pretty much like a gorilla, but with some obvious human traits—blue eyes, for example, or a more functional opposing thumb—perhaps the inventor has created something that resembles, approaches, or, in the awkward language of an ambivalent patent examiner, "embraces" a human being. And if the gorilla grows up with an ability to use its larynx, tongue, and lips well enough to articulate a few simple words, "I love you," for example, one might even grant the invention person-hood.

Partly in anticipation of human/animal chimeras, legal scholars and bioethicists have already begun to challenge the strictly human view of personhood. "Personhood doesn't come from coded sequences," asserts University of Pennsylvania bioethicist Arthur Caplan, "nor does it require human anatomy. It's really defined by what goes on in a brain. A dolphin, for example, could conceivably acquire enough additional intelligence to warrant personhood."

What, then, for "human rights"? And what rights are left for, say, humans in persistent vegetative states who Caplan and a growing number of his peers say might one day be considered to have lost their personhood and thereby their right to life? Save the dolphin with its humanesque reasoning and kill the vegetative patients by harvesting their organs for sick humans, or even for a part-human chimera?

But the real issue here is not with chimps, dolphins, mice, or any of their rights to personhood, but with our notion of humanity and how it is challenged by chimeras, which threaten either to erase taboos we still embrace, like bestiality, or reintroduce practices we'd hopefully sloughed off, like slavery. Could one animal cell make a being suitable for ownership, forced labor, and medical experimentation, just as "one drop" of black blood once did?

What becomes of human empathy if there are chimeric, quasi-human "flesh robots" performing human tasks, or if there are two, three, four, or more genetically separate hominid species of self-conscious, intelligent, soulful beings on the planet, perhaps one being genetically enhanced to be stronger, faster, brighter than today's humans, with enough chromosomal conflicts to prevent crossbreeding with "lesser" humans?

Will genetically enhanced humans lay claim to the Bill of Rights and exclude all others, or write their own Bill of Superhuman Rights? Will they regard the gene poor much as many European whites once regarded dark-skinned humans—as lesser, subhuman or nonhuman? And will the gene poor, the "naturals," be forced by fate, law, or judicial decree to accept their genetic lot in life? And who will parent the chimeras?

These are questions that Stuart Newman believes must be answered before a patent is considered for a human/primate chimera, and before a future court is forced by chimeric technology to determine the legal status of pigs with brains powerful enough to render them self-conscious. So he continues to fight for approval to devise creatures he hopes are never made.

GINA KOLATA

Stem Cell Science Gets the Limelight; Now It Needs a Cure

FROM THE *NEW YORK TIMES*

Stem cell research has been in the news as much for its potential in treating disease as for its ethical implications. But perhaps a more important question is whether it will even work. Gina Kolata investigates.

At three laboratories around Boston, separated by a taxi ride of no more than ten or fifteen minutes, the world of stem cell research can be captured in all its complexity, promise, and diversity.

One of the labs focuses on cells taken from human embryos, another on cells from mice and fish, and a third from stem cells that have mysteriously survived in the adult body long after their original mission is over.

But while the work here and elsewhere has touched off a debate reaching into the presidential campaign, a tour through these labs shows that the progress of research is both greater and less than it seems from a distance.

One idea, the focus of about half the nation's stem cell research, involves studying stem cells that are naturally present in adults. Researchers have found such cells in a variety of tissues and organs and say they seem to be a part of the body's normal repair mechanism. There are no ethical issues in studying these cells, but the problem is in putting them to work to treat diseases. So far, no one has succeeded.

The other line of research, with stem cells from embryos, has a different obstacle. Although, in theory, the cells could be coaxed into developing into any of the body's specialized cells, so far scientists are still working on ways to direct their growth in the laboratory and they have not yet effectively cured diseases, even in animals.

The most progress with embryonic stem cells is in mice, where one group of researchers directed the cells to grow into a variety of blood cells, but not yet the ones they want. Another group directed mouse stem cells to grow into nerve cells and tried to use them to treat Parkinson's disease in mice. The nerve cells produced the missing chemical, dopamine, but not enough to cure the disease.

As the two lines of research proceed along parallel paths, researchers say it is far too soon to bet on which, if either, will yield cures first. "It's not either-or," said Dr. Diana Bianchi, chief of the division of medical genetics at Tufts New England Medical Center in Boston.

At the medical center, Dr. Bianchi says, her foray into the world of stem cell research involved a decade of discoveries so unexpected that despite her stellar reputation, colleagues at first looked askance.

Dr. Bianchi, who works in a lab stretched out along a narrow corridor of an old building that was once a garment factory, stumbled into the field when she was trying to find a new method of prenatal diagnosis.

She knew that a few fetal cells enter a woman's blood during pregnancy and hoped to extract those cells for prenatal diagnosis. That proved too difficult because there are so few fetal cells in maternal blood.

But then she discovered that the fetal cells do not disappear when a pregnancy ends. Instead, they remain in a woman's body for decades, perhaps indefinitely. And if a woman's tissues or organs are injured, fetal cells from her baby migrate there, divide and turn into the needed cell type, be it thyroid or liver, intestine or gallbladder, cervix or spleen.

She and her colleagues find fetal cells by looking for male cells in tissues and organs of women who have been pregnant with boys and showing that the cells' DNA matches that of the women's sons or, if the women had abortions, their male fetuses. (Cells from female fetuses also enter a woman's body, but it is quicker and easier to find the male cells by looking for cells with a Y chromosome, Dr. Bianchi says.)

One woman, for example, had hepatitis C, a viral infection. But when her liver repaired itself, it used cells that were not her own.

"Her entire liver was repopulated with male cells," Dr. Bianchi said.

Such findings astonished even Dr. Bianchi. But now, with publications in leading journals, few doubt her.

In theory, fetal cells lurking in a woman's body are the equivalent of a new source of stem cells and could be stimulated to treat diseases. But, Dr. Bianchi says, she does not yet know for sure that the cells are stem cells—she must isolate them and prove they can turn into any of the body's specialized cells—nor where the cells reside, or how, short of injury, to spur them to action.

A SHORT DISTANCE AWAY, Dr. Leonard Zon, the chief of stem cell research at Children's Hospital, and his colleague Dr. George Q. Daley are working with stem cells from embryos, using mice and zebra fish for now. They want to learn how to transform the stem cells into immature blood cells that will divide and replenish themselves.

Then, if they can apply their work to human embryonic stem cells, they want to use the cells instead of bone marrow transplants to treat patients with genetic disorders like sickle cell anemia, and inborn disorders of the immune system.

Dr. Zon treats children with these diseases, most of whom do not have a relative whose cells match theirs closely enough to serve as a bone marrow donor. He urgently wants to help.

But he is not there yet. So far, in research that stem cell investigators say is among the most promising in the field, Dr. Zon and Dr.

Daley have turned mouse embryonic stem cells into mouse blood cells. Those blood cells, however, are more mature than the ones they need, a particular type of early blood cells that can repopulate a patient's bone marrow and survive indefinitely. Ones that are more mature live out their lifespans and die within weeks.

They are also working with human embryonic stem cells, venturing into the most controversial area of stem cell work. Human embryonic stem cells are derived from human embryos, about a week old, and the only way to get the stem cells is to destroy the embryos.

Some human stem cells came from embryos that were donated by couples at fertility labs who had embryos left over after they decided their families were complete. Others came from embryos that were created to obtain stem cells; researchers paid women to donate eggs, fertilized them and let them grow to the stage where stem cells could be extracted.

The federal government has agreed to pay for research with human stem cells, but only for work with twenty-two lines; each line is the progeny of a single embryo. That restriction dates from August 9, 2001, when President Bush issued a directive saying the government would pay for research, but only with cell lines created before that date.

Dr. James F. Battey, director of the National Institute on Deafness and Other Communication Disorders and chairman of the National Institutes of Health's Stem Cell Task Force, said scientists were free to study other stem cell lines if they used private money. He understands the researchers' complaints that it would be better if the government paid for work on more lines, but, he said, as far as the federal government is concerned, "the argument isn't solely about science."

"What the president has already said on multiple occasions is that he is committed to the notion that taxpayers' money should not be used to encourage the destruction of human embryos," Dr. Battey said. "This is a White House policy."

And, he said, "it is not based solely on the needs of the scientific community."

But Dr. Zon said being able to work on more human stem cell lines could help the research.

"When you are trying to do research, you look for every advantage you can," he said. "Some embryonic stem cell lines make particular tissues better than others."

Some, for example, might more easily turn into blood cells, and others might more easily grow into nerve cells, but there is no way to know whether there is a better stem cell line for a particular cell type without trying as many as possible, Dr. Zon said. "You would want to find the line that makes the tissue you are studying."

ACROSS THE RIVER in Cambridge, in the basement of a biology building on Harvard's campus, a small group of scientists works in a two-room lab on the site of a former machine shop. Among their goals is to plunge into one of the most controversial areas of stem cell research—creating human embryos by cloning and obtaining stem cells from those embryos.

An embryo created by cloning would be an exact genetic match of the person whose cells were used to make it. Its stem cells and any mature cells derived from those stem cells would exactly match the cells in the person's body, making them perfect replacement cells.

One of the four part-time researchers at the Harvard lab, Dr. Kevin Eggan, learned to clone mice as a Ph.D. student at MIT and, he said, the group is seeking approval from Harvard's ethics committee to try to start the cloning process with human cells.

The federal government forbids the use of its money to pay for such research, but this lab, directed by Dr. Douglas A. Melton, a Harvard developmental biologist, takes no federal money. Instead, the work is paid for by the Howard Hughes Foundation, the Juvenile Diabetes Foundation and the Naomi Berry Diabetes Center of Columbia University.

Cloning, however, can be onerous. In February 2004, researchers in South Korea announced that they had gotten stem cells from human embryos they created by cloning, but they began with 176 human eggs and ended up with one embryo that yielded stem cells.

Dr. Eggan, though, is not after replacement cells. His goal in cloning is to understand what goes wrong in a disease like Alzheimer's, Parkinson's or diabetes.

Dr. Melton gave an example. Suppose he had stem cells that were exact matches of fifty patients with Parkinson's disease and directed them to grow in the laboratory into nerve cells of the type that die in the disease. He could then ask when, and why, the cells die.

"Do they all show a defect at the same stage? If so, that would mean there is a common cause, like a flat tire. Or maybe each one breaks down in a different way. Is there one way to get Parkinson's, or fifty ways?" Dr. Melton asked.

"We could use that information to do drug screening," he added, possibly finding ways to prevent the nerve cell death.

For now, though, the Harvard lab is becoming a supplier to the world of its own seventeen lines of human embryonic stem cells, created without cloning, and made from 286 frozen embryos created by in vitro fertilization.

Meanwhile the national debate over the use of human embryonic stem cells goes on.

While many Americans say in polls that they favor using these cells, many others have strong moral objections. Creating and destroying a human embryo to obtain stem cells, they say, is ethically unacceptable, and doing research on human embryonic stem cell lines that are already in existence does not right the wrong.

It is "a kind of after-the-fact cooperation with this destruction," said Richard Doerflinger, deputy director of pro-life activities for the United States Conference of Catholic Bishops.

The challenge for scientists in the midst of a fierce political debate, many say, is to be realistic about how hard it is to develop treatments.

Dr. Battey lists some of the challenges ahead: getting the cells to develop into exactly the adult cells that are needed, demonstrating that the adult cells can survive, preventing rejection and controlling cell growth.

Such issues, Dr. Battey said, "need to be addressed in animal models before any thoughtful person would go into humans."

PHILIP ALCABES

The Bioterrorism Scare

FROM *THE AMERICAN SCHOLAR*

What will claim more lives? A man-made smallpox epidemic? An outbreak of a powerful new influenza strain? Or the yearly toll from cancer? The public-health scholar Philip Alcabes makes the compelling—and contrarian—argument that our preoccupation with bioterrorism may be blinding us to more pressing threats to public health.

Since the fall of 2001, when America embarked on a "war on terrorism" and federal officials started warning us about the next plague, here are some events that have *not* happened: a U.S. epidemic of sudden acute respiratory syndrome (SARS), a widespread anthrax outbreak, any smallpox attack, the discovery of hard evidence of a biological weapons program in Iraq. Yet the talk about "biopreparedness" continues. Some people in Washington want the Centers for Disease Control and Prevention to be transferred from the Department of Health and Human Services to Homeland Security. There is even a professional journal, *Biosecurity and Bioterrorism*, devoted to learned discussions of the topic. Is it sound public policy to rush to protect the country against the threat of attack with germs that could cause an epidemic? Does the *bio* in *biosecurity* mean that we should turn our public health into a matter of civil defense? Or have we Americans been sold a bill of goods?

Throughout history, the responses to both actual communicable disease and the threat of it have been guided by the metaphor of the stranger as the spreader of contagion. Allegations that epidemic disease was caused by foreigners are ancient. Thucydides reported that his contemporaries, in the fifth century BC, attributed the Plague of Athens to Ethiopians. Later on, such thinking was refined to impute causation specifically to *enemy* foreigners. Many European and American authors still believe that the Black Death entered western Europe after the Mongols, besieging the city of Kefe (or Kaffa, now Feodosiya in Ukraine) in 1346, catapulted into the European-held city the corpses of comrades who had died of plague. When the siege was lifted, the theory goes, the disease reached Genoa aboard ships. Although Kefe clearly experienced plague in 1346, the story of its source is almost certainly apocryphal: *Xenopsylla cheopis*, the flea that usually carries the plague bacillus, has affinity only for warm bodies and will desert a corpse within hours of death. But the myth's persistence attests to the readiness of some, even centuries later, to believe that epidemics were caused by enemies.

Similarly, Londoners believed that the Great Plague of 1665 was brought by the Dutch, with whom England was then at war (echoes of that belief appear in both Defoe's *Journal of the Plague Year* and Samuel Pepys's *Diary*), although there is no evidence that plague in fact came to England from Holland. The great influenza pandemic of 1918–19, which killed more than twenty million people—possibly as many as forty million—in a world at war, was attributed to various enemies. In fact, the name by which we remember this epidemic, "Spanish Flu," seems to have been a compromise acceptable to the warring parties (Spain was neutral in the war). And in a telling instance, very shortly after Germany broke its pact with the Soviet Union and invaded its former ally, Hitler used the term *Pestherd*—"plague focus"—to refer to Russia, accusing the Soviets of infecting

Europe with Jewish "bacilli." It was the metaphor of foreign culpability for disease turned inside out, Hitler imputing pathogenic properties to Jews and creating a new enemy by alleging that it was Russia that had spawned them. More recently, Tanzanians attributed the high death rate from AIDS in their Kagera province to HIV infections brought by Idi Amin's Ugandan troops, who crossed the border into Tanzania in the late 1970s. "It made sense, then," Laurie Garrett writes in *The Coming Plague*, "to assume that the new disease came from old enemies." When West Nile encephalitis made its Western hemisphere debut in New York in 1999, epidemiologists received calls from the FBI; they (and the CIA, too) were concerned that it might be the work of foreign bioterrorists.

The Stranger Spreading Germs shows up often in literature and film. Alessandro Manzoni's 1821 novel, *I Promessi Sposi*, attributes the advance of plague in 1629 to the German army, then campaigning through the valleys of northern Italy in the Thirty Years' War. In the novel, foreigners, particularly the French, come under suspicion in plague-ridden Milan, where they are accused of "daubing" plague-inducing substances on walls or sprinkling plague powders on the streets; those found guilty of such felonious behavior are tortured to death. In F. W. Murnau's classic silent film *Nosferatu*, plague arrives in Bremen by sea, brought from the East by the odious Other, the undead Nosferatu. In 1922, just after the carnage of World War I and the far greater mortality of the Spanish Flu pandemic, Murnau depicted Evil as the man who was beyond death, the inscrutable and indomitable being who brings pestilence from the benighted *Ausland*.

SOMETIMES, OF COURSE, epidemics *have* come from the enemy foreigner. Clearly, Cortés was able to conquer Mexico because of smallpox. The disease appeared among the Taino on Hispaniola in 1518, brought by Spaniards who had colonized the island; later, it would contribute to the Taino's extinction. By 1519 smallpox was in Cuba. Cortés, who was secretary to the governor of Cuba, left to take

Tenochtitlán from the Aztec chief Montezuma; either Cortés's Spanish troops or those of Narváez, who led a relief expedition against the Aztecs after they repulsed Cortés's initial sallies, brought smallpox to the Aztecs. The disease, to which the Aztecs were immunologically naïve, so diminished their numbers that Cortés had only to finish them off. Smallpox thence spread southward, killing the Incan emperor Huayna Capac and then his son in 1524–25, and plunging their people into civil war. Francisco Pizarro had little more to do to vanquish the Incas than march into Cuzco.

It is unlikely that the Spaniards infected the American natives deliberately—and the Aztecs, at least, seemed to interpret the devastation as evidence of divine disfavor, not treachery on the part of Spain. But in the French and Indian War, in the early 1760s, smallpox does seem to have been spread deliberately. During the conflict known as Pontiac's Rebellion, Sir Jeffrey Amherst, the British commanding general, approved a plan to distribute smallpox-contaminated blankets "to innoculate [sic] the Indians" besieging Fort Pitt, at the fork of the Ohio River. That epidemic smallpox occurred in the Ohio Valley at the time is an established fact.

Around the same time, the Polish army considered producing cannon shells filled with the saliva of rabid dogs, in an attempt to poison the air the enemy breathed. Later, during World War I, the German biological warfare program sought to create animal epidemics—*epizootics*, in the lingo of epidemiology—that would diminish their enemies' ability to fight. German agents deliberately infected their neutral trading partners' livestock and animal feed with the agents of glanders (principally an equine disease) and anthrax. Romanian sheep were infected with both microbes in 1916 before export to Russia; more than two hundred Argentine mules intended as dray animals for Allied forces died after being inoculated with both bacteria; French cavalry horses were infected with glanders; and attempts were made to contaminate animal feed in the United States. During World War II, the Third Reich avowedly eschewed use of biological warfare, but one report holds that Colorado beetles were dropped by German airplanes on potato crops in southern England. The U.S. Army claims that the retreating Wehrmacht contaminated a reservoir in Bohemia with sewage, presumably to produce disease in the advancing Soviet army.

The best documented, and most successful, deliberately caused human epidemic was set by the infamous Unit 731 of the Japanese Imperial Army, stationed in conquered China during World War II. In addition to numerous heinous medical "experiments"—tortures, really—the unit dropped plague-carrying fleas on eleven Chinese towns. The number of Chinese who died of plague was probably about seven hundred. Unit 731 also produced cholera. In fact, Japanese soldiers, whom the unit had failed to warn or prepare, died in large numbers after entering Chinese areas seeded with *Vibrio cholerae*. Some historians put the total number of deaths attributable to the unit's intentional contaminations, including others involving anthrax and typhoid, in the thousands.

In connection with humans deliberately causing epidemics, the smallpox germ, *Variola virus,* is the one we hear most about. It enjoyed a five-hundred-year career as a natural epidemic pathogen, from roughly the late fifteenth century until the late twentieth. Then smallpox was eradicated from the earth. *Variola* killed several hundred million people in the first half of the twentieth century alone—a public health menace to be reckoned with. No doubt because of its fearsome reputation, it is the subject of a great number of speculative scenarios about how it might be resurrected as an epidemic scourge.

In fact, though, none of those scenarios is even remotely likely. First, not many people have access to viable smallpox stocks. Second, the disease that *Variola virus* produces is fairly easy to diagnose. Third, vaccination will prevent disease even in already-infected contacts of smallpox cases, and vaccine stocks are reasonably large nowadays. The hubbub about smallpox has had the effect of sharpening physicians' diagnostic skills (indeed, so much so that instances of overdiagnosis produce false alarms) and expanding the supply of available vaccine. At this point, standard public health procedures, including case diagnosis, contact investigation, and immunization of possibly infected individuals, would be adequate to prevent an outbreak in the unlikely event that some individuals were deliberately infected.

Anthrax is the second most popular topic of bioterrorism conversation. We have seen intentional anthrax infection—the much-ballyhooed postal anthrax events that took place in the fall of 2001. Three characteristics of that outbreak are of note: very few people became ill; very, *very* few died; and it was almost certainly not produced by a stranger.

Environmental studies in mailrooms indicated that many hundreds of people were probably exposed to anthrax spores that fall, yet only twenty-two people got sick. And of those twenty-two, half had cutaneous anthrax, the rarely-life-threatening skin form of the disease. Only five died. In the jargon of epidemiology, anthrax turned out to be neither very infectious nor very pathogenic. That experience should tell us that spraying anthrax spores from crop dusters or releasing them from aerosol cans into the subway is highly unlikely to make many people ill.

Speculation about subway attacks stems from a real event in March 1995, when the Japanese religious cult Aum Shinrikyo released the nerve toxin sarin in the Tokyo subway system. Twelve people died. Two subsequent attempts to release toxins in the Tokyo subways were foiled. Note that Aum was using a gas, which does not have to be sprayed; it diffuses by itself. This is not how germs are disseminated, and it is a distinction worth bearing in mind. And even that ignores the more central question of *likelihood*. Large-scale poisonings are not easy to carry out well.

The light death toll from mailed anthrax was a result of the low pathogenicity of the bacteria—half the cases were not pulmonary and were therefore unlikely to be fatal—and the comparative treatability of anthrax disease once detected. Do five deaths constitute a public health crisis? Along with his colleagues, Victor Sidel, Distinguished Professor of Social Medicine at Albert Einstein College of Medicine in New York, has noted that a fraction of our nation's expenditure on biopreparedness would pay for effective treatment of tuberculosis for all of the two million people who get TB each year in India, thereby preventing close to half a million deaths a year. Half a million deaths because commonly available antibiotics are not affordable—now *there's* a public health problem.

The other microbe that is on the lists of virtually all the bioterrorism watchers is the plague bacillus. It is true that Unit 731 produced plague outbreaks in China by dropping infected fleas on towns. But at that time plague was a recurring problem in Asia: a ferocious epidemic struck Manchuria in 1910, and another occurred in 1921. (It still is a problem: a large outbreak caused many deaths in India as recently as 1994.) By contrast, despite the presence of *Yersinia pestis,* the plague bacterium, in wild rodents in the Western Hemisphere, there has never been an extensive epidemic of human plague in this country. Even when plague epidemics moved out of Asia through much of South America, circa 1900, the United States saw only a small outbreak in San Francisco's Chinatown. The reason is not that Americans are immune to plague; it is that the urban arrangements that we have been accustomed to for the past two hundred years are inhospitable to the rat-flea-bacillus ecosystem. Such reforms as garbage removal, pest control, and better housing explain why plague disappeared from eastern Europe in the early 1700s and has never troubled us seriously here. Since epidemics of plague are unlikely, should we then worry that terrorists will produce isolated cases? Perhaps, but garden-variety antibiotics are very effective at treating the disease and interrupting transmission. There is no potential for the next catastrophe there.

Other pathogens have been mentioned as possible bioweapons— for example, the agents of tularemia, botulism, and Q fever. These organisms are not generally transmitted from person to person, so they carry little or no outbreak potential. Hemorrhagic fever viruses are sometimes transmitted by mosquitoes or by the bite of infected animals. It has never been shown that they can be manipulated into transportable weapons and then elude standard mosquito- and animal-control programs.

———

ALL IN ALL, there is little evidence that terrorists are more likely, or better able, to use microbes as part of their armamentarium than ever before. If there *were* evidence, would editorialists in the nation's most prestigious medical journal need to argue, as they did in the context of the purported smallpox threat, that public health decisions should rely on "theoretical data"? Consider the ratio of known success to attempts at bioterrorism.

Jessica Stern reports in *The Ultimate Terrorists* that the Aum Shinrikyo cult drove three trucks set up to spray botulinum toxin through Tokyo in 1990, but no cases of botulism resulted. Judith Miller, William Broad, and Stephen Engelberg report in their book, *Germs*, that Aum sprayed anthrax spores from the roof of its building in Tokyo in 1993. This apparently killed some birds, but no humans got sick. Aum members also reportedly tried to procure Ebola virus in the 1990s from what was then Zaire, but if they were able to get the virus, they were unable to produce any cases of Ebola.

In the 1960s a Japanese researcher purposely contaminated food with *Salmonella typhi*, producing outbreaks of typhoid fever and dysentery, but no deaths. In 1970 four Canadian students got sick after eating food that had been deliberately contaminated with pig ringworm ova. A neo-Nazi group in the United States stockpiled the typhoid bacterium in 1972, with the intent of contaminating the water supplies of midwestern cities. They failed. In 1984 members of the Rajneeshee cult contaminated ten Oregon salad bars with *Salmonella typhimurium*, which causes diarrhea. Federal investigators located 751 cases of salmonellosis, but no deaths occurred. In 1996 twelve people in Texas developed dysentery after eating doughnuts or muffins purposely contaminated with shigella bacteria by a disgruntled lab worker. Again, no deaths.

Presumably, these initiatives represent only a small subset—the known ones—of all attempts to cause mayhem by deliberately introducing germs into a population. Indeed, one systematic study uncovered twenty-nine such attempts to use germs to harm others, although a large proportion seemed to have been perpetrated by single individuals whose aims were against other individuals rather than the population. No doubt there are still more instances, unknown because they were unsuccessful. And yet we can enumerate on the fingers of one hand the deadly epidemics that have in fact been caused deliberately. In the creation of epidemics, the gap between intention and the deed itself is a wide one.

Americans began looking at new infectious diseases as a grave social menace in the early 1990s, and that anxiety resonated loudly after the publication of Laurie Garrett's *Coming Plague* in 1994. In a sense, the relatively recent worry about bioterrorism is a spin-off of the past decade's concern about the "coming plague." In reality, the intensity of even naturally occurring new epidemics never matches that anticipation.

Our most recent experience with so-called emerging infections has been SARS. It appeared suddenly in southern China in late 2002, the causative virus probably having entered the human population through people who had substantial contact with domestic animals. Although SARS affected more than eight thousand people worldwide and killed more than seven hundred, it was a negligible problem in the United States (eight cases, no deaths) and produced no more than a mild epidemic in most other countries: despite the high case-fatality ratio (people diagnosed with SARS had about one chance in ten of dying from the disease), only in China, Hong Kong, Singapore, and Canada were there more than five SARS deaths. Only nineteen countries saw more than a single SARS case. In all the affected areas, the outbreak was brought under control within about six months of its onset.

Two aspects of the SARS experience are important. First, though it is easy to acquire the virus by inhaling respiratory secretions from a SARS sufferer, it also turns out to be easy to prevent or control an outbreak. The key is to use standard infectious-disease-control measures, including case finding and reporting, active surveillance at points of entry to the country, isolation of possible cases, and recommendations against travel to heavily affected regions. None of these measures requires cutting-edge technology; all have been used in controlling communicable disease for well over a century.

Second, SARS made news partly because it was *not* the expected epidemic. For six months before the advent of SARS, Americans had been worrying very publicly about smallpox. Urged toward apprehension by the federal government, we had alarmed ourselves about the possibility that the long-defunct disease would be reborn in the hands of bioterrorists. The administration made ready, in mid-2002, to vaccinate half a million armed-services personnel and half a million health care workers. The former plan it came close to accomplishing; the latter was abandoned because so many health care workers refused to show up for vaccination. All of it repeatedly made the headlines and the evening news reports. Yet what happened in the end was not smallpox, or smallpox prevention; it was SARS. Had our faces not already been turned toward epidemic disease and our anxieties about infection elevated, had the news hours not been hungry for *new* news after a month of relentless coverage of the Iraq war, SARS might not have made such big headlines.

The lesson we should learn from our experience with SARS is that if we are vigilant about spotting new disease outbreaks and equally vigilant about applying public health programs to curtail their spread, we can limit them, although we cannot ward them off completely. We cannot make life risk-free. Had we dealt with AIDS and West Nile encephalitis the way we dealt with SARS (which, of course, may return in the future and therefore requires continued vigilance), their course might have been different. West Nile, ending only its fifth season in the United States as I write, is already virtually a national epidemic; AIDS went national within six or seven years of its appearance. But in its initial season, 1999, in New York City, West Nile virus caused forty-six cases of encephalitis and seven deaths: not negligible, but a public health problem more minor than Lyme disease (and far less extensive, in New York, than asthma or lead poisoning). AIDS began with a handful of cases—there were only a few hundred in 1981, the first year it was recognized—and the United States might have kept the toll fairly small had we had the political nerve to do something about it at the time. The point is that, at the outset, none of these "plagues" were cataclysmic. Epidemics do not work that way. Errant microbes do not find their way into the human ecosystem and wipe out most of the population unannounced. *The Andromeda Strain* is fantasy.

THE CRYSTAL BALL with which we divine epidemic mayhem is no clearer now than it used to be, and no clearer than the vision with which we try to foresee the coming plague. No one can dispute, after the events of September 11, 2001, that some people wish us harm. However, that harm is not likely to come from bioterrorism. To worry that the Middle Easterner, the Arab, or any Muslim—however the Stranger is configured—will use germs to attack us would be to pretend that we can indeed foresee great epidemics. For two reasons, that is certainly not the case.

First, humans, even ill-tempered and badly behaved humans, have never been able to use germs or weapons to the terrible degree of mortal effect that nature always has been able to use germs. Just four communicable diseases—malaria, smallpox, AIDS, and tuberculosis—killed well over half a *billion* people in the twentieth century, or about ten times the combined tolls of World Wars I and II, history's bloodiest conflicts. Even today, with good vaccines and effective antibiotics to stop them, infectious diseases kill about ten million people each year.

The worst catastrophes the world has seen have been not the genocides, however gruesome, but the cataclysmic disease outbreaks. When a few million people are killed by design with Zyklon B, the machete, or the machine gun, it is a horror and an outrage; it shakes our moral faith. But the Black Death killed a third of Europe's population in just four years in the mid-1300s. Smallpox wiped out entire tribes of American natives after Europeans arrived in the 1500s. Plague killed over fifty thousand in Moscow alone in just a few months in 1771. The Spanish Flu killed between twenty and forty million in sixteen months in 1918–19.

And that suddenness is the second point. Prevision is of little help against epidemic disasters. Each of the disasters I just mentioned, and every other great epidemic of history, was unimaginable until the moment it began. But neither is prescience necessary: each epidemic, even the ones that turned out to be most terrible, began slowly, percolated a while, and could have been stopped with conventional public health responses had anyone acted in time. SARS reminded us of that.

Whether or not the Stranger is an enemy avowedly bent on terror, casting him in the role of microbial evildoer has the perilous effect of distracting us from realizing two truths: the disheartening one that the epidemic crystal ball is always cloudy, and the uncomfortable one that it is usually social circumstances that make epidemics possible and public health funding that stops them.

Worrying about the germ-bearing Stranger, we forgo the upkeep of a workaday public health apparatus in favor of fabricating modern wonders. The CDC now operates what it calls a "war room." From there, it can coordinate activities around SARS, West Nile virus, and other infections, as well as bioterrorism—if it can be found—using high-tech communications equipment. The war room is in an "undisclosed location." Once, CDC officials knew they were running a public health agency. Now, apparently, they must act as if they are in charge of national defense.

After the U.S. Department of Health and Human Services scotched the 2002 plan to vaccinate health care workers, the CDC announced it would continue its effort to vaccinate police officers and firefighters, despite numerous reasons to stop: several deaths directly attributable to the vaccine; several more possibly attributable; evidence that people who might be predisposed to heart disease (something like 10 or 15 percent of the middle-aged population) can be harmed by the vaccination; certainty that people with HIV infection (a sizable percentage of the adult population in some big-city neighborhoods) must not be vaccinated; and of course the complete absence of natural smallpox infection anywhere in the world for the past twenty-five years.

Federal grant money, under President Bush's multimillion-dollar Project BioShield program, has been allocated to technologic innovation for bioterrorism prevention. By 2003, according to *The Chronicle of Higher Education,* the National Institutes of Health were supporting almost seventy extramural research projects on anthrax alone. The NIH has funded two new National Biocontainment Laboratories and new facilities at Regional Biocontainment Laboratories, most at major universities, to the tune of $360 million in start-up costs. The University of Pittsburgh just lured the top staff of the Center for Civilian Biodefense Strategies away from Johns Hopkins by offering to set up a Center for Biosecurity with a $12 million endowment. The University of South Florida recently received $5 million in federal grant money for its Center for Biological Defense, which has projects like "Photocatalytic Air Disinfection" and "Aquatic Real-time Monitoring System (ARMS) for Bioterrorism Events." And Auburn University received a million-dollar federal biopreparedness grant for something called a Canine Detection Center.

When Harvard received a $1.2 million federal grant in 2002 to set up a program for detecting "events possibly related to bioterrorism" by electronically linking twenty million patient-care records from around the country (an endeavor called syndromic surveillance), the grant was a mere drop in the bucket: Congress soon allotted $420 million to Homeland Security for a larger linked health-monitoring network. A consortium led by the New York Academy of Medicine then developed software for syndromic surveillance. The new software allows public health officials to monitor what are called "aberrant clusters health events"—translation: more than the expected number of cases of some symptom that might be related to a disease that might be produced by an organism that might be in the possession of terrorists. Syndromic surveillance works fine if someone knows what to look for. But it is of no help at all with the unexpected. When there really was bioterrorism in the United States—the anthrax attacks in fall 2001—linked databases were useless; it took a smart clinician to figure out that anthrax was around, and old-fashioned shoe-leather epidemiology quickly worked out which people had been affected. West Nile virus, ditto. A friend who is an official of a local health department tells me that the syndromic surveillance experts find it works well for predicting the first influenza outbreak each year. But anyone's grandmother can predict the first influenza outbreak where I live, in the New York City area, since the symptoms are always the same and flu always starts here in the two weeks preceding Thanksgiving. For all its meager productivity, the syndromic surveillance software also throws out plenty of false-positive "clusters" that waste investigators' time. It is our new and expensive white elephant, justified by the fear of evildoers and germs.

It all sounds a little phantasmagoric.

In *The Exposé of 1935,* Walter Benjamin identified certain phantasmagorias of modern life as *Wunschbilder,* "wish symbols," that seek to "transfigure the . . . inadequacies in the social organization." Benjamin was concerned with phantasmagorias as magical images, as he put it in his *Passagen-Werk,* "residues of a dream world," revelations of unfulfilled wishes. In this sense we might ask whether the many projects of the BioShield—the electronic anthrax detectors, the hyperdatabanks, the supersecure biohazard-level-4 labs, the men and women in full-body protective suits, the urban-evacuation exercises, whatever it is they are doing with dogs at Auburn University, and so on—are emblems of some shared desire to feel that all the effort America expends on technology development protects us from *something.* We might ask whether the something is the Stranger Spreading Germs. And we must ask what the cost is.

THE CORE ISSUE HERE is that the Stranger Spreading Germs is a metaphor, and largely an empty one. Bioterrorism is not a public health problem, and will not become one. The next plague, whatever it is, will not decimate us unheralded. In signing the contract on biopreparedness, we have bought a confection, a defense against the chimeric stranger with the metaphorical germs.

And the costs of buy-in? When our public health "leaders" corroborate government rhetoric about bioterrorists by reassuring us that our state or municipal health department is ready for any smallpox, anthrax, or plague attack, they legitimate both their own efforts and the standing of their offices. The planned result is that we will not question spending tax dollars so those officials can continue to defend us, even when it means closing down municipal clinics or shortchanging programs for the poor, and even if such biodefense is not what we most need or want.

The biopreparedness campaign goes to work. It discredits the simple logic of public health. Lose the distinction between the minuscule risk of dying in an intentional outbreak and the millionfold-higher chance of dying in a natural pandemic, it says. Ignore the hundred-fold-higher-still chance of dying of cancer or heart disease. Defund the prenatal care clinics, the chest clinics, the exercise and cancer screening and lead abatement programs. Ignore the lessons of history: forget that human attempts to create epidemics have almost always failed, and dismiss the repeated ability of a well-funded public health apparatus to control epidemic disease with time-tested measures. Just think about germs, and tremble.

The lesson of history that we ignore at our peril is this: Nobody can tell us how the next epidemic will happen. New germs come and go, epidemics wax and wane, but national catastrophes happen rarely. And when they do, it is never because of a stranger spreading germs. Anyone who promises certain protection from the next plague is selling us a bill of goods.

LAURIE GARRETT

The Hidden Dragon

FROM *SEED*

While not at high levels now, AIDS is poised to explode in Vietnam. The government has a plan to deal with it, but they need the world's help. Laurie Garrett measures whether action by the global community will prevent a Southeast Asian pandemic.

An amber full moon loomed large over Hanoi, casting a glow over otherwise dimly lit streets. Tens of thousands of scooters and motorcycles zigzagged chaotically along Ly Thuong Kiet Street, slicing through the steamy, sweltering air. Entire families were crammed onto individual scooters, gangs of teens stacked themselves onto 90 cc bikes, and the comparatively minimal number of cars and trucks simply rammed their ways into the scooter mobs, seemingly oblivious to the lives at stake. The traffic din was deafening, gasoline exhaust redolent—and yet, paradoxically, the entire aesthetic was charming. That's Vietnam: Against all evidence of poverty, pollution, political chicanery, and repression, it charms the heck out of you with its French Colonial buildings, young people's naïve candor, lemongrass aroma, and studied simplicity.

As I made my way up moonlit Ly Thuong Kiet, sweat raining down my forehead and back, I noticed a line of Vietnamese men standing on tiptoe, craning to see something on the third floor of the Melia Hotel. I joined them, and it took me a few moments to realize with astonishment what had caught their eyes. The entire ballroom of the luxury hotel was done up in star-spangled banners. Red, white, and blue confetti streamed from the ceiling, well-dressed Americans danced to hip-hop with comparatively dainty Vietnamese partners, and big-screen video projectors showed Hollywood stars cavorting in Los Angeles. The Vietnamese men gawking from the street were stunned. They had no idea they were witnessing a party to celebrate American independence from the British Empire. As it wasn't actually the Fourth of July yet but a couple of days before, few of the celebrants probably cared what the fete's excuse was: In Hanoi, any rationale for a party will do.

I was taken aback. The last thing I'd ever expected to see was an American flag party in Hanoi. Still, I had to admit that the Vietnamese, whether they knew it or not, had momentary cause to celebrate the United States of America.

All day long I had been in meetings with health officials from all over the world, responding to the staggering news that the White House had selected Vietnam to receive millions of dollars under the President's Emergency Plan for AIDS Relief, or PEPFAR. With less than three months to lay the groundwork for the barrage of PEPFAR initiatives, the country's local United States, Vietnamese, and United Nations health workers were scrambling madly to figure out how to process millions of dollars effectively—and without fueling Vietnam's legendary corruption. The roughly $30 million expected to pour in before the end of September 2004 will more than double Vietnam's annual AIDS prevention and treatment budget.

Vietnam doesn't really have a huge AIDS epidemic—yet. HIV is there, of course, but the numbers are small and the epidemic is quite young, probably having started in 1997–98. Officially, about 215,000 people are infected right now in Vietnam: An astonishing 62 percent of them are under thirty years of age. Local scientists estimate that the true infection rate is one out of every eighty-one Vietnamese—a far cry from the one-out-of-three rates seen in much of Africa, but well above the rates seen in the United States and Western Europe.

VIETNAM HAS ALL THE ELEMENTS in place for an AIDS explosion. In the alleyways and bars of those sweltering Hanoi streets, some young people are "chasing the dragon"—smoking wild concoctions of heroin, speed, and anything else they can get their hands on. The United Nations Programme on HIV/AIDS (UNAIDS) recently did a calculation that blew away Communist Party leaders in Hanoi, showing that roughly one out of every seventy-five households nationwide has a family member suffering HIV infection or AIDS. The best estimate, accepted by the Communist Party, is that about forty to fifty Vietnamese become newly infected every day, accounting for roughly 0.5 percent of the daily worldwide burden of new infections. The Vietnamese leaders adore numbers and statistics—this goes with their style of communist management—and that household-infection figure sparked a real shake-up in the party earlier this year, I was repeatedly told.

Drug use is rampant all over Southeast Asia. These kids aren't smoking pot and popping Ecstasy on the weekends, igniting outbreaks of friend-fondling like their American counterparts. No way. This is an insane scene in which highly calculated drug traffickers are hooking a whole generation on whacked-out combinations of benzodiazepines, methamphetamines, heroin, and whatever else is around. And the manufacturers aren't small-time bikers making meth in their Iowa basements. Hell, no. Like everything in Asia, the scale is staggering, and these drugs and mixes of powders and pills are made in massive factories located chiefly in Burma and southern China. They have been flooding Southeast Asia with variations on this hellish crap for several years, but Vietnam didn't witness the damage until after it opened its borders to global trade in 1997—and drug trade unfortunately followed.

Local narcotics experts told me the drug traffickers have taken a calculated bead on Ho Chi Minh City in particular, winning over teenagers with a clever campaign called "The Pink Pill." Attractively packaged bright pink pills are flooding the marketplace, advertised by street dealers as fun highs. Lab analysis shows they are harmless caffeine pills, akin to No-Doz in the States. That might seem odd—except getting thirteen-year-olds used to the idea of buying illegal pills that make them jittery and speedy opens the door to a customer base that is starting to join the ranks of slightly older methamphetamine and narcotics users. Now Vietnam, having survived two decades of war with the French and another decade of ultimately victorious battle against the U.S. military, finds itself smack-dab in the middle of the global war on drugs.

Openness and global trade have also brought about a resurrection of the war-era prostitution scene in Vietnam, which is going full tilt in Hanoi and Ho Chi Minh City and spreading nationwide. The commercial sex-worker trade, as public health people politely call it, is fueling HIV spread all over Asia, along with explosions in syphilis, gonorrhea, chlamydia, hepatitis B, and a host of other sexually transmitted diseases. The Vietnamese sex trade is unusual for the region, however, in that its customers are overwhelmingly local guys, rather than foreigners. It's easy to see how widespread the prostitution scene is: Just look for signs that say "KaraOke," a deliberate miswriting of "karaoke bar." The inserted large *O* stands for—well, use your imagination.

Add one more element to the mix, according to regional health officials: age. Asia—especially south of the Chinese border—is experiencing an enormous baby boom. In Hanoi, it's hard to spot anybody over fifty years of age, and under-thirty-year-olds seem to control the streets day and night. Privately, Communist Party officials told me that they are losing control of even their own children as the influx of outside culture, hip-hop, drugs, movies, and fashion is capturing the hearts, minds, veins, and lifeblood of their youth. According to U.N. agencies, a third of Vietnam's population is under fifteen years of age, and the median age of the Vietnamese is merely twenty-four years.

These risks for HIV aren't the primary reasons the White House selected Vietnam as its fifteenth PEPFAR recipient, all of the other countries being in Africa and the Caribbean. There was political pressure coming from many quarters, officials told me, to add an Asian country to the agenda. With the Fifteen International AIDS Conference scheduled for Bangkok in July, the White House hoped to announce an Asian designee before the July 11 opening of the world's largest AIDS gathering. The fastest-growing Asian epidemic is assumed by many scientists to be in Burma (a.k.a. Myanmar, if one chooses to adopt the ruthless government's renaming), but that military dictatorship is on the U.S. boycott list. Many nations are trying to help Burma in its fight against HIV, but they do so through a specially controlled, pooled fund that is managed by a consortium of Western governments and U.N. agencies. By agreement, none of them donates directly to the military regime. Cambodia has a terrible epidemic unfolding, but it's a very hard country to work in because of its acute poverty and political instability. More than one percent of the general populations in Burma, Thailand, and Cambodia are now HIV positive—3 percent in Cambodia—according to Tim Brown, a leading epidemiology consultant to UNAIDS. Together, China and India represent the greatest global HIV threat, as combined they have roughly 2.25 billion human beings. Even 2 percent prevalences in those countries could have devastating impacts on the global resources to fight disease. But officials told me Vietnam was selected for two reasons: It seemed easier because its epidemic was comparatively small and new, and it has obvious symbolic value for Americans.

Regardless of why Vietnam was chosen, the newness of its epidemic and the well-organized nature of its government offer an opportunity to test just how far the global AIDS community has come in prevention strategies. Can HIV be stopped in its tracks in Vietnam by using the scientific and social toolbox developed over the last twenty years?

The Vietnamese government thinks it has no choice: It must pull out the stops, using every prevention tool science has invented. Recently the Communist Party released a national strategy for AIDS control, expressing a vision of action going all the way out to the year

2020. The official objective is "to control the HIV prevalence rate among the general population to below 0.3 percent by 2010," and to nearly eradicate the virus by 2020. This fairly awesome target will be achieved, the national-strategy document declares, by the "Central Committee of Vietnam Fatherland Front and its member organizations," which will "mobilize the participation of the entire population in HIV/AIDS prevention and control, integrate HIV/AIDS prevention, and control activities into mass agitation movements," and so on and so forth, in true Marxist-Leninist fashion. Rhetorical as this all sounds, Vietnamese leaders I spoke to are utterly earnest, and U.N. officials in Hanoi believe them.

ONE STEAMY AFTERNOON I prowled around Hanoi's massive public facility, Bach Mai Hospital, where the bulk of the northern Vietnamese AIDS cases are warehoused. It's the sort of medical center communists the world over loved in the late twentieth century—the kind I have seen all over the former Soviet Union, Eastern Europe, and in African countries that were heavily influenced by the old Soviet Union. Bach Mai is huge, with thousands of hospital beds, most in close proximity to others and offering no patient privacy. One of the buildings is nearly twenty-five thousand square feet. Increasingly those beds are filled with AIDS patients, though Bach Mai officials refused to provide numbers or allow me to visit their AIDS wards. The hospital is composed of several buildings, spanning some five urban block equivalents. It's a city unto itself, with cafes, stores, Buddhist temples, multiple pharmacies, and a thriving street-vendor scene. The underpaid doctors and nurses earn paltry sums for their forty-hour workweeks, generally less than fifteen dollars per month, and often go missing as they supplement their incomes with private practices. Health care is a pay-as-you-go affair for most Vietnamese, and getting sick is an enormously costly event. Bach Mai is not a place one wants to inhabit for long.

I was most interested in seeing whatever memorial might be in place to commemorate Bach Mai's lowest point, in 1972, when the

U.S. Air Force dropped four sorties of bombs on the hospital, obliterating the entire downtown Hanoi medical complex, killing dozens of patients and health care workers. My older brother, Banning, was there at the time and filmed the devastation for CBS's *Evening News*. As I wandered the halls of the vast, rebuilt facility, I recalled Walter Cronkite's introduction to my brother's footage, saying it had not been shot by a CBS employee but seemed to depict a horrible set of attacks carried out by U.S. military personnel against a civilian medical facility. For the Garrett family, Bach Mai Hospital has always had a special historical resonance.

After walking a seemingly endless series of halls, seeing and smelling the patients who idled away long, steamy hours on steel beds, I stumbled into the Administration Building. I noticed a photo display ahead of me, started toward it, and was stopped by a military officer. After a half hour of negotiations involving several hospital officials and Vietnamese military personnel, I was finally permitted to step further into the lobby to view the memorial photo display. It proved a depressing exhibit of faded old photos poorly mounted on easels and walls, depicting the terrible condition that U.S. bombing had left the hospital in. In less time than it had taken to negotiate the privilege of seeing the display, I stepped back out into the blinding midday sun, feeling the lingering tensions of U.S.-Vietnamese relations. History does not conveniently erase itself.

By the time U.S. Global AIDS Coordinator Randall Tobias arrived in Hanoi this summer, shortly after America's Independence Day, the Vietnamese Communist Party had decided to completely reverse its policy of involuntary incarceration of prostitutes and drug users to stem the tide of HIV—and was looking for new solutions. Last year the trend was toward so-called 05/06 Centers, named after a Party fiat declaring the need to detain drug users. In theory, those young people placed in the 05/06 centers went into long-term detention willingly—even volunteered out of desperate need to detox and clean up their lives. In truth, local experts told me, 05/06 detainees have been rounded up by police, committed by their parents and employers, fingered by their neighbors, and targeted by their school and university administrators. Nobody knows how many young people have been locked up under 05/06, but estimates range to the hundreds of thousands, perhaps even a million. Some have spent less than a month in a center; many face two years of confinement.

The 05/06 centers, still in operation, are run by the Ministry of Labor, War Invalids, and Social Affairs (MOLWISA), a central government agency that seems truly desperate to find humane and lasting treatments for drug use. But each center is managed by local authorities, with regional Communist Party bosses playing key roles. The result is a patchwork of programs that United Nations officials told me range from quite decent job-training and drug-rehabilitation centers to horrible jail facilities that rival the worst juvenile prisons seen in the Western world. MOLWISA knows the 05/06 system isn't working well and is desperate for new ideas. Despite the draconian 05/06 system, experts in Hanoi told me that a third of the nation's drug users are now infected with HIV, and 60 percent of drug users in Haiphong carry HIV.

Meanwhile, time is running out. Recent government and U.N. studies show that prostitutes in Ho Chi Minh City are starting to use methamphetamines to give themselves greater stamina on the job. Some of the young meth users are not only mixing heroin and meth but, even worse, are now old enough to stagger in their stupors to the prostitutes, creating a potent potential for mixing the two risk groups and sparking wider spread of HIV.

In Thailand, there has been little cross-spread of HIV between sex workers and drug users because the two communities are distinctly separate. Not so in Vietnam, according to numerous new studies. In Hanoi, for example, 57 percent of men who inject drugs say that they frequently see prostitutes; 17 percent of the prostitutes admit to injecting drugs. Molecular analysis of the HIV strains circulating in Hanoi confirm that Vietnam's epidemic is not heading in a Thai direction, because the two chief risk groups strongly overlap and share viral types.

All over Asia, from southern Indonesia to northern China, this volatile mix of methamphetamines, heroin, and prostitution threatens to blow up the currently low prevalence epidemics in the region. At the Bangkok AIDS meeting, Karen Stanecki, a prominent demographer working for UNAIDS, said that Vietnam was one of a handful of Southeast Asian countries wherein "HIV infection among injecting drug users has shot from near zero to 50 percent or more in just a few years."

Stanecki and Tim Brown worked with a team of experts this year to produce a comprehensive analysis of the Asian AIDS epidemic. They concluded that the numbers of infected in the region remain small but that Asia's risks are acute, in large part because the vast majority of drug users and prostitutes in cities like Hanoi are under twenty-nine years of age. Youth brings recklessness; when combined with sex-for-money and potent drugs, the situation, they said, is a powderkeg. The conservative Asia Development Bank (ADB) predicts an additional ten million people in the region will become infected over the next six years unless drastic prevention steps are taken. ADB says the impact on regional economies will be "catastrophic."

A World Health Organization (WHO) official in Hanoi told me that Vietnam's HIV problem is still trivial compared to the astounding numbers of people who are slaughtered and maimed in motorcycle accidents. As he talked I felt a shiver, recalling having seen a young couple on a motor scooter that morning, neither of them wearing helmets. The woman was sitting sidesaddle, her high heels dangling as the scooter zigzagged at a terrifying speed through enormous herds of bikes and cars. She was clutching a newborn baby in one arm and holding onto her man's waist with the other, and I wanted to reach out and save that baby. The WHO official was right: At this moment in Vietnam's history, creating a semblance of road safety would save a lot more lives than developing an HIV program.

But HIV is an epidemic: Without interventions of some sort, it will grow and spread and one day overwhelm those traffic accident figures. Vietnam knows this and wants help.

But what can PEPFAR do? In a speech a week later to the Fifteenth International AIDS Conference in Bangkok, Tobias acknowledged the limitations on him, noting that the U.S. government program might not be able to help in Vietnam in the ways it can in Africa.

"Unlike most of our other focus nations," Tobias said, "Vietnam's HIV/AIDS epidemic has been fueled by intravenous drug use. As we develop our country plan, we will be looking at ways to educate those who inject drugs about the added risk of HIV/AIDS and exploring means to support drug-abuse prevention and treatment."

Okay, but let's be clear: In addition to the crazy methamphetamine scene, Vietnam does have a heroin-injection problem, and addressing that would mean methadone and clean needles to drug users. But Congress doesn't want PEPFAR to spend U.S. taxpayers' money buying either. Vietnam does not have an HIV epidemic in the mainstream heterosexual population, such as Africa does, so messages about being faithful to one's mate or abstaining from sex simply aren't relevant. The methamphetamine problem means that hundreds, maybe thousands, of young people are disinhibited periodically, and then likely to indulge in casual sex with partners or prostitutes: They need condoms. But, again, this is strictly limited under PEPFAR.

Ironically, the U.S. government has its hands tied on issues about which the Vietnamese Communist Party has decided to take action. In its official AIDS-control plan, the Communist Party calls upon its members "to carry out comprehensive harm-reduction intervention programs, including the clean-syringe and needle-exchange program and 100 percent condom-use program, in areas where the number of injecting-drug users and sex workers as well as the HIV prevalence rate is high."

The Vietnam dilemma is an Asian paradox: Across the world's most populous continent, PEPFAR would run into the same contradictions between the need to slow the pandemic and Washington's abhorrence of drug users and sexual frolickers. Before anti-Bush readers jump to the conclusion that allowing for condoms and needles would solve the PEPFAR paradox, however, they need to take a hard look at the methamphetamine, poly-drug scene and ask themselves just what harm reduction is for these people. It's not an easy question to answer, and science has no pat solutions. There is no methadone equivalent for methamphetamines, and the only way to smooth out a cold turkey is with calming agents, like benzodiazepines. But the clever, nefarious dealers of Asia are mixing those agents into the speed concoctions, and many young users are addicted to benzodiazepines.

This stuff simply is not easy.

WE MIGHT HAVE A LIVELY DEBATE in scientific circles about how best to tackle this, but the current political climate in Washington makes that all but impossible. On July 14, 2004, when U.S. Global AIDS Coordinator Randall Tobias spoke at the Fifteenth International AIDS Conference in Bangkok, he was, predictably, booed and shouted at by conference participants. The top political official in the Department of Health and Human Services (HHS), Bill Steiger, set limits on how many federal employees were even allowed to attend the conference—just fifty—and ordered that CDC and other agency scientists who got nongovernment funds to attend could not do so. The American Medical Association (AMA) sponsored a briefing in Bangkok before the conference started, but the key scientist responsible for research on preventing mother-to-child transmission of HIV in the United States was not allowed to attend. The AMA said it offered to pay the CDC scientist's way, but the HHS still blocked his travel. Throughout the conference there were CDC and NIH presentations and posters full of data—all without the presence of the primary scientists. This meant that all sorts of idea sharing and interaction between researchers from all over the world could not take place. Standing in front of a poster presentation (about nine thousand data sets were posted, rather than presented orally in Bangkok) on HIV infection among Sudanese militias, I was frustrated, boiling over with questions. In light of the situation in Darfur, it would have been nice to have had the CDC researcher standing beside his poster, as is usually the case, offering additional insights to his work. But nobody was there. The poster simply had to speak—poorly, as it turns out—for itself.

Steiger and his fellow travelers in the Bush administration are squelching debate on sexual and drug-use issues in many, many ways. Over the summer they added two startling new layers of blockades. First, Steiger ordered that the WHO could no longer directly request a given U.S. scientist by name for assistance in epidemics, technical advisory panels, or work in the Geneva headquarters of the agency. Instead, requests must now be filed in a generic sense: "We need some guy who knows something about genital herpes for a committee meeting next month." Steiger and his office will then select a politically appropriate—and scientifically skilled?—individual to go to Geneva. Top officials in Geneva told me that they have only worked this way with a handful of countries over the entire fifty-plus years of WHO's existence: the Soviet Union, China, India, and a few authoritarian regimes. Indeed, WHO is officially using its old Soviet Union policies now in dealing with the United States.

And Steiger and Tobias have led a charge against the Global Fund to Fight AIDS, Tuberculosis, and Malaria, the key financing mechanism for new assaults on disease worldwide. In 2002, Republicans in Congress insisted the United States could contribute no more than 30 percent of the Global Fund's annual revenues, with the remainder being donated by the poor countries themselves, Europe, Canada, and Japan. Recently the Global Fund has had a great deal of success in pushing European investment, and Fund director Richard Feachem went to Steiger and Tobias hat in hand, saying the United States needed to give more to hit its 30 percent for fiscal year 2005. Steiger said no, insisting that, according to Congress, a U.S. fiscal year commences October first. Therefore, the United States will only match contributions made by Europe and other partners over nine months' time, not twelve. And over the nine months, he insists, Europe failed to donate sufficiently to force a hike in U.S. commitments.

So much for empirically based governance.

In the 1970s, the United States discovered that tiny Vietnam was more than it could handle. Perhaps PEPFAR will discover the same thing in the twenty-first century, forced to bend its own rules and rethink its policies. It is ironic, in the extreme, that the Communist Party of Vietnam currently displays more flexibility in its AIDS policies than does the government of the United States of America.

The Mop-Up

FROM *THE NEW YORKER*

> *Programs to eradicate infectious diseases have not always succeded,*
> *with smallpox as a famous exception. Currently, the World Health*
> *Organization is in the closing stages of a campaign to stamp out*
> *polio. The surgeon and writer Atul Gawande accompanies a World*
> *Health Organization team to a poor region in India, where they try*
> *to limit the damage from what might be the last few cases of polio*
> *in the world.*

The index case was an eleven-month-old boy with thick black hair his mother liked to comb forward so that the bangs rimmed his round face. His family lives in the southern Indian state of Karnataka, in a village called Upparahalla, along the Tungabhadra River. Dry mountains of teetering rocks can be seen in three directions from the village. It has no running water and little electricity. The boy's mother is illiterate; the father can read only road signs. They are farm laborers, and they live with their three children in a single-room hut of thatch and mud. But the children are well nourished. The mother wears gold and silver earrings. Once in a while, they travel.

In April 2003 the family took a trip north to see relatives. Shortly after they returned, on May 1, the boy developed high fevers and racking bouts of nausea and vomiting. His parents took him to a nearby clinic, where a doctor gave him an antibiotic injection. Two days later, the fevers subsided, but he became unable to move either of his legs. In a panic, the parents took him back to the doctor, who sent him to the district hospital in Bellary, about forty miles away. As the day progressed, the weakness spread through the boy's body. His breathing grew shallow and labored. He lay flat and motionless in his hospital cot.

A doctor at the hospital, following standard procedure in cases of sudden childhood paralysis, phoned a surveillance medical officer with the World Health Organization in Bangalore, the capital of Karnataka. The medical officer made sure that proper cultures were taken and sent to a national laboratory in Mumbai. On June 24, the laboratory results came back. A young technical officer with the WHO in New Delhi got the call; it was a confirmed case of polio, a disease thought to have been eliminated from southern India, and it set off an alarm.

The World Health Organization is in the sixteenth year of a campaign to eradicate polio from the world. If the campaign succeeds, it may be mankind's single most ambitious accomplishment. International organizations are fond of grand-sounding pledges to rid the planet of this or that menace. Such pledges make the organizations feel that they are doing something important. But they nearly always fail. The world is too vast and too various to submit to dictates from on high.

A handful of serious attempts have been made to eliminate individual diseases from the world. In 1909 the newly established Rockefeller Foundation launched the first global eradication campaign, an effort to end hookworm disease, using anti-helminthic drugs, in fifty-two countries. It didn't work. Today, a billion people—a sixth of the world's population—are infected with hookworm, an intestinal parasite that feeds on human blood. A seventeen-year campaign against yellow fever, led by the Rockefeller Foundation and the United States armed services, had to be abandoned in 1932 when yellow fever was found to have a reservoir outside human beings. (The yellow-fever virus persists in mosquitoes' eggs.) In 1955 the WHO and UNICEF began a campaign to end yaws, an infectious disease causing painful, purulent skin ulcers; workers screened a hundred and sixty million people in sixty-one countries for the disease, and treated every case they found with penicillin. A dozen years later, the campaign was dropped when it turned out that silent, subclinical infections were continuing to propagate the disease. Billions of dollars were spent in the fifties and sixties to eradicate malaria; today the disease afflicts more than three hundred million people a year.

In the course of a century, the only successful attempt at disease eradication has been the battle against smallpox—a mammoth undertaking that was, however, decidedly simpler than the campaign against polio. Smallpox, with its distinctive blisters and vesicles, could be easily and quickly identified; the moment a case appeared, a team could be dispatched to immunize everyone the victim might have come into contact with. That strategy, known as "ring immunization," eradicated the disease by 1979. Polio infections are far harder to identify. For every person who is paralyzed, between two hundred and a thousand infected people come down with little more than a stomach flu—and they remain silently contagious for several weeks after the symptoms abate. Nor is every case of childhood paralysis polio; and it usually takes a couple of weeks for stool specimens to be obtained, delivered to a laboratory, and properly tested. By the time one case has been identified, scores more people have been infected. As a result, the area targeted for polio immunization must be far larger than that for smallpox. And, whereas people needed to be vaccinated against smallpox only once for immediate protection, a single dose of polio vaccine does not always take—children with diarrheal illnesses tend to pass the vaccine straight through—and so a repeat round of immunization is required within four to six weeks. In logistical terms, it's the difference between extinguishing a candle flame and putting out a forest fire.

Despite all these obstacles, the campaign against polio has made immense progress. In 1988 more than three hundred and fifty thousand people developed paralytic polio, and at least seventy million were infected with the virus. In 2002 only 1,919 cases were identified. The whole of the Americas, Europe, and the western Pacific, along with nearly all of Africa and Asia, are currently free of the disease.

India is the one country where polio infections have substantially increased in recent years, from a low of 268 in 2001 to 1,600 in 2002, when it accounted for four-fifths of the world's remaining cases. With its vast population, areas of severe poverty, and varied cultures and geography, it is the place where the campaign against polio is at greatest risk of failure. The outbreak in 2002 was a serious setback, but it was at least confined to a handful of northern states. Now a boy in south India had the disease—Karnataka's first case in almost three years—and if it wasn't checked there it would blaze across the country all over again.

ON JUNE 25, less than twenty-four hours after the report of the Karnataka polio case came in, Sunil Bahl, a WHO physician and technical officer in the Delhi office, sent an e-mail to key people at the WHO, at UNICEF, and in the Indian government. It was his job to provide the initial assessment of the facts on the ground. "The case is in an area that has a history of being the worst in Karnataka," he wrote; it had the most polio cases in the early years of the campaign, and poor routines of immunization. "Risk of establishment of virus in the area high, unless quick wide and strong measures in the form of a wide mop-up are taken." A "mop-up" is WHO lingo for a targeted campaign to immunize all susceptible children surrounding a new case. It's what is done in an area that has been rendered polio-free but is facing a new infection that threatens to bring the disease back. The campaigns are highly targeted, and are carried out rapidly, in just three days, to insure that children are not missed and to make it easier to recruit volunteers.

Sunil Bahl sent around a map of the proposed area for the mop-up operation, an area covering fifty thousand square miles. Working around the summer holidays and festivals, government officials selected July 27 for the start of the first immunization round. The second round would follow a month later. Brian Wheeler, a thirty-five-year-old Texan who is the chief operations officer for the WHO's polio program in India, explained the logistics to me. The Indian government would have to recruit and organize teams of medical workers and volunteers, he said. They would have to be trained in how to administer the vaccine, and provided with transportation, vaccine, and insulated coolers and ice packs to keep the vaccine cold. And they would have to fan out and vaccinate every child under five years of age. Anything less than 90 percent coverage of the target population would be considered a failure.

I asked him how many people that would involve.

He checked his budget sheet. The plan, he said, was to employ thirty-seven thousand vaccinators and four thousand health care supervisors, rent two thousand vehicles, supply more than eighteen thousand insulated vaccine carriers, and have the workers go door to door to vaccinate 4.2 million children. In three days.

POLIO IS CAUSED BY AN INTESTINAL VIRUS; the virus must be ingested to bring about an infection. Once inside the gut, it passes through the lining and takes up residence in nearby lymph nodes. There it multiplies, produces fevers and stomach upset, and passes back into the feces. Those infected can contaminate their clothing, bathing sites, even supplies of drinking water, and thereby spread the disease. (The virus can survive as long as sixty days outside the body.)

Poliovirus infects only a few kinds of nerve cells, but what it infects it destroys. In the most dreaded cases, the virus spreads from the bloodstream into the neurons of the brain stem, the cells that allow you to breathe and swallow. To stay alive, a person has to be fed through a tube and ventilated by machine. The nerve cells most commonly attacked, though, are the anterior horn cells of the spinal cord, which control the arms, the legs, and the abdominal muscles. Often, so many neurons are destroyed that muscle function is eliminated altogether. Tendon reflexes disappear. Limbs hang limp and useless.

The first effective vaccine for polio was introduced in 1955, after the largest clinical trial in history. (Jonas Salk's vaccine, made from killed poliovirus, was given to four hundred and forty thousand children; two hundred and ten thousand received a placebo injection, and more than a million served as unvaccinated controls.) Five years later, Albert Sabin published the results of an alternative polio vaccine he had used in an immunization campaign in Toluca, Mexico, a city of a hundred thousand people, where a polio outbreak was in progress. His was an oral vaccine, easier to administer than Salk's injected one. It was also a live vaccine, containing weakened but intact poliovirus, and so it could produce not only immunity but also a mild contagious infection that would spread the immunity to others. In just four days, Sabin's team managed to vaccinate more than 80 percent of the children under the age of eleven—twenty-six thousand children in all. It was a blitzkrieg assault. Within weeks, polio had disappeared from the city.

This approach, Sabin argued, could be used to eliminate polio from entire countries, even the world. Curiously, the only person in the West who took him up on the idea was Fidel Castro. In 1962 Castro's Committee for the Defense of the Revolution organized 82,366 local committees to carry out a succession of week-long, house-to-house national immunization campaigns using the Sabin vaccine. In 1963 only one case of polio occurred in Cuba.

Despite those results, Sabin's grand idea did not catch on until 1985, when the Pan American Health Organization launched an initiative to eradicate polio from the Americas. (Six years later, Luis Fermin Tenorio, a two-year-old boy in the town of Pichinaki, Peru, became the last polio victim in the Americas.) In 1988, spurred by the campaign's growing success, the WHO committed itself to eradicating polio from the world. That year, Rotary International pledged a quarter of a billion dollars for the effort (it has since provided $350 million more). UNICEF agreed to organize the worldwide production and distribution of vaccine. And the United States made the campaign one of the CDC's core initiatives, supplying both expertise and considerable additional funding.

The centerpiece of the effort has been what are called national immunization days—three-day periods when all children under five in a country are immunized, regardless of whether they have received immunization before. In one week in 1997, two hundred and fifty million children were vaccinated simultaneously in China, India, Bhutan, Pakistan, Bangladesh, Thailand, Vietnam, and Burma. In each of the past three years, national immunization days have reached more than half a billion children—almost a tenth of the world's population. Through such efforts—and a reliable network of monitors to detect outbreaks—the WHO campaign has brought the incidence of polio in the world to less than one percent of what it used to be.

The striking thing is that the WHO doesn't really have the authority to do any of this. It can't tell governments what to do. It hires no vaccinators, distributes no vaccine. It is a small Geneva bureaucracy run by several hundred international delegates whose annual votes tell the organization what to do but not how to do it. In India, a nation of a billion people, the WHO employs two hundred and fifty physicians around the country to work full time on polio surveillance. The only substantial resource that the WHO has cultivated is information and expertise. I didn't understand how this could suffice. Then I went to Karnataka.

FOR THE THREE DAYS OF THE MOP-UP, I travelled through Karnataka with Pankaj Bhatnagar, a WHO pediatrician, whose job was to see that the operation was properly executed. He is in his forties, with a slight paunch and an easy, genial manner. This can be a tricky business, he explained as we waited in Delhi for our flight to Karnataka. The WHO provides much of the money for mop-up operations. UNICEF provides the vaccines. Rotary of India prints the banners and advocates locally for the cause. But the operation itself is run by government health officials: they are the ones who must hire the thousands of vaccinators, train them properly, and send them from house to house.

We took a plane to Bangalore, then travelled eight hours overnight by train to Bellary, a crowded, dusty town that is the district seat for Upparahalla. At a small, strange hotel there (it had a safari theme), Pankaj convened the members of his team over breakfast. To monitor the immunization of four million children, he had just four people: three young medical officers and himself. They were the only ones available who spoke Kanada, the local language. The medical officers finished their breakfast of *idli* and *dosa,* and lit up cigarettes (in India, it seems, half the doctors who work in public health smoke), and then Pankaj asked for a status report.

Since the index case was identified, he was told, four more cases of confirmed polio had appeared in the region, and four "hot" cases were awaiting confirmatory testing. Of the thirteen districts targeted for mop-ups, Bellary accounted for all but one of the cases.

"Then we must concentrate our monitoring in this district," Pankaj said. "This is now the place with the most intense transmission of polio in the world." Another doctor pulled out some figures on the area. Bellary district, he told Pankaj, has a population of 2,965,459, with 542 villages and nine urban towns. Fifty-two percent of the males and 74 percent of the females are illiterate. There are just ninety-nine doctors in the public health system. He turned to a map. The polio cases, he said, were clustered in a triangle of villages around Siriguppa, a small, slum-ridden town about forty miles away.

Pankaj made his assignments. For the mop-up, he would cover at least Upparahalla; a village called Sirigere, where polio had appeared; the two urban areas with hot cases; and a mine in Chitradurga, where vaccinators might have particular difficulties gaining entry, because the housing was on the property of a private company. He divided the others among the remaining villages and asked them to follow up behind him for a second check in Upparahalla and the urban areas.

The group then split up. By eight-thirty in the morning, Pankaj and I were on the road.

We had a rented four-wheel-drive Toyota and a betel-nut-chewing driver who waited until we were an hour down a pitted road to tell us that the battery was dead. Whenever the engine was turned off, he said, we'd need to push-start the car. Pankaj thought this was funny.

The terrain outside the windows was baked by the hot sun, and the hills were desert-lizard brown. The monsoon had failed to come this year. Only the few fields that had drip irrigation looked green. It took us about two hours to travel the thirty-five miles to Sirigere, a village of mud-walled huts jammed up against one another. There was garbage in the alleyways, and dusty-faced children were playing everywhere. Pankaj had the driver stop at a group of dwellings seemingly at random. Marked in chalk on each door was a number, a "P," and that day's date. The number was the house number. The "P" meant that the vaccinators had come, identified all the children under the age of five who lived in the house, and vaccinated them—that very day, according to the date marked. Pankaj took out a pad of paper and strode over to one of the huts. He asked the young woman at the door how many children lived there. One, she said. He asked to see the child. When she found him, Pankaj took his hand and noted the black ink mark on the nail bed of his little finger—it's how the vaccinators tag the children who have received polio drops. Was any other child in the fields? Away at a relative's? No, she said. He asked if her boy had received routine immunizations before today? No, she said. Had she heard about the polio case in town? She had. Had she heard about the vaccination team before the workers arrived at the door? She had not. He thanked her and wrote all the information down on a form before moving on.

So far, the workers had done their job, Pankaj said, several houses later. But he was disturbed that no one knew the vaccinators were coming that day. In addition to banners (we'd seen a couple hanging as we came into the village), they were supposed to use "miking" to reach the illiterate—auto-rickshaws with loudspeakers playing tapes announcing the upcoming campaign. Without that warning, some people would turn the vaccinators away.

Going around to a few more huts, we bumped into a vaccination team—a social-welfare worker wearing sandals, a blue sari, and a flower in her hair, and a younger, college-student volunteer with a flower in her hair, too, and a square blue cold box of vaccine slung over her shoulder. They were standing in front of a hut they'd marked with an "X" instead of a "P"—the woman of the house had said that three children lived there, but one was absent and could not be vaccinated. Pankaj asked the vaccinators to open their cold box. He checked the freezer packs inside—still frozen, despite the heat. He inspected the individual vaccine vials—still fresh. There was a gray-and-white target sign on each vial. Did they know what it meant? That the vaccine was still good, they said. What does it look like when the vaccine expires? The white inside the target turns gray or black, they said. Right answer. Pankaj moved on.

We went to the home of the village's recent polio case. The girl was eighteen months old and silent. The mother, pregnant and with a three-year-old boy clinging to her side, laid her down on her back so that we could examine her. Neither leg would move. Lifting each one, I felt no resistance in the child's hips, her knees, her ankles. Only four weeks had passed since she was stricken. She almost certainly was still contagious.

Pankaj found three children visiting the house. He checked each of their hands. None had received polio drops yet.

WE GAVE THE FOUR-WHEEL-DRIVE A PUSH and made our way to Sirigere's primary health center, a few miles outside the village. It was a drab, unpainted, three-room concrete building. The center's medical officer met us at the door. About forty years old, with ironed slacks, a buttoned short-sleeve shirt, and the only college education in the area, he seemed eager to have our company. He offered tea and tried to make some small talk. But Pankaj was all business. "May I see your microplan?" he asked before we had even sat down. He was referring to the block-by-block plan drawn up by each local officer. It is the key to how the operation is organized.

The medical officer's microplan was a sheaf of ragged paper, with marker-drawn maps and pencilled-in tables. The first page said that he had recruited twenty-two teams of two vaccinators each to cover a population of 34,144 people. "How do you know this population estimate is right?" Pankaj asked. The officer replied that he'd done a house-to-house survey. Pankaj looked at the map—the villages in the area were spread out over more than ten miles. "How do you distribute the vaccine to the vaccinators who are far away?" By vehicle, the officer said. "How many vehicles do you have?" Two, he said. "What are the vehicles?" One was an ambulance. The other was a rented car. "And how does the supervisor get out to the field?" There was a pause. The officer shuffled through the microplan. More silence. He did not know.

Pankaj went on. Twenty-two teams would require about a hundred ice packs per day. "Why did you budget for only a hundred and fifty ice packs?" We are freezing them overnight for the next day, the officer explained. "Where?" He showed Pankaj his deep freezer. Pankaj opened it up and pulled out the thermometer, which revealed that the temperature was above freezing. The electricity goes out, the officer explained. "What is your plan for that?" He had a generator. But when pressed to show it he was forced to admit that it wasn't really working, either.

Pankaj is not a physically imposing man. He has a boyish mop of thick black hair, parted almost down the center, and sometimes it sticks up. He has programmed his cell phone to play the James Bond theme when it rings. When we're driving, he points out the monkeys we pass. He makes jokes. He laughs with his head tilted back. But in the field his demeanor is grave and taciturn. He doesn't tell people if their answers are good or bad. He keeps everyone on edge. I had an impulse to tell the medical officer that he was doing fine. But Pankaj seemed to make a point of saying nothing to fill the silences.

Later, in Siriguppa, where two of the hot cases had appeared, we walked the neighborhoods with another medical officer. Siriguppa is a dense, urbanized town of windowless concrete-block tenements, rusting corrugated-metal lean-tos, and some forty-three thousand people. We had to fight our way through narrow streets crowded with water buffalo, motorcycles, braying goats, and fruit sellers. There was electricity here, I noticed, running through wires that drooped from tilting utility poles, which were scattered around like crooked teeth, and the sound of televisions poured out from some of the houses we visited.

The two hot cases, we found, were in a small Muslim enclave that had sprouted up a few months earlier. Going door to door, Pankaj learned that almost none of the enclave's children had received routine immunizations. Some of the families seemed suspicious of us, answering questions tersely or trying to avoid us altogether. We found one boy whom the vaccinators had missed. The previous year, rumors had circulated widely among Muslims that the Indian government was giving different drops to their male children in order to make them infertile. The rumors were thought to have been quashed by an education campaign and greater Muslim involvement in the immunization program. But one had to wonder.

Walking with a local doctor and a vaccination team through a village called Balkundi, we came to the home of a small, pretty woman, who had rings on her toes and a baby held loosely on her hip. Another child, a boy of about three, stood nearby, staring at our little crowd. Neither child had been vaccinated, so Pankaj asked if we could give them the polio drops. No, she said. She did not appear angry or afraid. Pankaj asked if she knew that a case of polio had appeared in her neighborhood. Yes, she said. But she still didn't want the drops given. Why? She would not say. Pankaj said okay, thanked her for her time, and moved on to the next house.

"That's it?" I asked.

"Yes," he said.

The local doctor had stayed behind, however, and when we looked back he was shouting at the mother: "Are you stupid? Your children will become paralyzed. They will die."

It was the one time I saw Pankaj angry. He walked back and confronted the doctor. "Why are you shouting?" Pankaj demanded.

"Before, she was listening, at least. But now? She's not going to listen anymore."

"She is illiterate!" the doctor shot back, embarrassed to be rebuked so openly. "She doesn't know what is right for her child!"

"What does that matter?" Pankaj replied. "Your shouting doesn't help anything. And neither does a story going around that we are forcing drops on people."

So far, few were refusing the drops, and that was good enough, he told me later. A single nasty rumor could destroy the whole operation.

ONE DIFFICULT QUESTION CAME UP REPEATEDLY—from local doctors, from villagers, from workers trudging house to house. The question was: Why? Why this huge polio campaign when what we need is—fill in the blank here—clean water (diarrheal illness kills five hundred thousand Indian children per year), better nutrition (half of children under three have stunted growth), working septic systems (which would help prevent polio as well as other diseases), irrigation (so a single rainless season would not impoverish farming families)? We saw neighborhoods that had had outbreaks of malaria, tuberculosis, cholera. But no one important had come to visit in years. Now one case of polio occurs and the infantry marches in?

There are some stock answers. We can do it all, goes one. We can eradicate polio and do better on the other fronts. In reality, though, choices are made. For that whole week, for instance, doctors in northern Karnataka had all but shut down their primary health clinics in order to carry out the polio-vaccination work.

Pankaj relies on a somewhat more persuasive line of argument: that ending polio is, in fact, worth diverting resources for. In one village, I watched a resident demand to know why the government and the WHO weren't combating malnutrition there instead. There was only so much they could do, Pankaj said. "And, if you're starving, becoming paralyzed certainly isn't going to help."

Still, you could make the same claim for almost any human problem that you decide to tackle—blindness or cancer or, for that matter, kidney stones. ("If you're starving, kidney pain certainly isn't going to help.") So far in the sixteen-year polio campaign, an estimated five million cases of paralytic polio have been averted, and that is an extraordinary achievement. But the campaign has already cost three billion dollars, more than six hundred dollars a case. To put that in perspective, the Indian government's total budget for health care came to four dollars per person in 2003.

Even if the campaign succeeds in the eradication of polio, it is entirely possible that more lives would be saved in the future if the money were spent on, say, building proper sewage systems or improving basic health services. What's more, success is by no means assured. The WHO has had to extend its target date for eradication from 2000 to 2002, and then to 2005. In these final years of the campaign, more and more money is spent chasing the few hundred cases that keep popping up. A certain weariness settles in. Around twenty-four million children are born in India each year, creating a new pool of potential polio victims the size of Venezuela's entire population. Just to stay caught up, a mammoth campaign to immunize every child under the age of five has been planned for this month. Stopping the very last case of polio, one official told me, might cost as much as two hundred million dollars. The truth is, no cost-benefit calculus will tell us whether all that money is well spent.

There is, nonetheless, a kind of greatness in the elimination of a terrible disease. We as a civilization have few things we can accomplish of genuinely lasting significance for mankind: we have built no pyramids, no Great Walls to stand for thousands of years. It is, instead, through medicine that we may create our enduring monument. The eradication of smallpox and now, perhaps, polio will stand as our pyramids.

But this means we must actually get down to that final polio case. Otherwise, the efforts of the hundreds of thousands of volunteers, the billions spent will have amounted to nothing, and perhaps worse than nothing. To fail at this venture would put into question the very ideal of eradication.

Beneath the ideal is the gruellingly unglamorous and uncertain work. But there is a system, and it has eradicated polio in countries with far worse conditions than I was seeing in India—for example, in Bangladesh, in Vietnam, in Rwanda, in Zimbabwe. Polio was eradicated from Angola in the midst of a civil war. An outbreak in Kandahar in 2002 was halted by a WHO-led mop-up operation despite the Afghan war. New mop-ups are now under way in northern Nigeria, where an outbreak recently appeared and spread into neighboring countries. In India, Pankaj told me, there have been campaigns on camels in the Thar Desert of Rajasthan, in jeeps among the tribal communities of the Jharkhand forests, on power boats through flooded regions of Assam and Meghalaya, on Navy cruisers travelling to remote islands in the Bay of Bengal. We covered about a thousand miles in our Toyota in the three days of going town to town. Outside the village of Balkundi, we came upon several makeshift shanties for migrant laborers, about four miles apart and not on any maps. When we checked the children, though, they all had the vaccinators' ink marks on their pinkies. At Chitradurga, we found the mines in decay, but state officials had made sure that the company gave the vaccinators access to the workers' compound. With some searching, we discovered a few children here and there. Every one of them had received the vaccine, too.

By the end of the mop-up, UNICEF officials had bought more than five million doses of fresh vaccine and distributed them through the thirteen districts. They had blanketed television, radio, and local newspapers with public-service announcements. Rotary of India had printed and delivered twenty-five thousand banners, sixty thousand posters, and more than six hundred and fifty thousand handbills. And four million of the targeted 4.2 million children had been vaccinated.

India has had just thirty-six cases of polio in the last three months of 2003 because of such efforts. There have been almost no cases during that time in northern India, a region that had more than a thousand cases in 2002. Pankaj and his colleagues believe that they're finally close to their goal of eradication. And as India goes, so, we can expect, will the world.

PANKAJ SAYS THAT HE HAS SEEN more than a thousand cases of polio in his career as a pediatrician. When we drove through the villages and towns, he could pick out polio victims at a glance. They were everywhere, I began to realize: the beggar with two emaciated legs folded under him, rolling by on a wooden cart; the man dragging his leg like a club down the street; the passerby with a contracted arm tucked against his side.

On the second day of the mop-up, we visited Upparahalla, the village where the Karnataka outbreak had started. The first, index case of polio was now a fourteen-month-old boy with a healthy, almost muscular thickness about his upper body; after the first few days of his infection, his breathing had returned to normal. But when his mother put him down on his stomach you could see that his legs were withered. With the exercises the nurses had taught her to do with him, he had regained enough movement in his left leg to be able to crawl, but his right leg dragged limply behind him.

Making our way around the open sewage, the mud-covered pigs, the cows resting curled up like cats with their heads on their hooves, we found the neighbor girl who had come down with polio after the boy. She was eighteen months old, with a big worried face, perfect white teeth, and short spiky hair. She was wearing small gold earrings and a yellow-and-brown checked dress. She squirmed in her mother's arms, but her legs only dangled beneath her dress. Her mother wore an impassive expression as she stood before us in the sun, holding her paralyzed child. Pankaj gently asked her if the girl had had polio drops before—perhaps she'd received the vaccine but it had not taken. The mother said that a health worker had come around with polio drops a few weeks before her daughter became sick. But she had heard from other villagers that children were getting fevers from the drops. So she refused the vaccination. A look of profound sadness now swept over her. She had not understood, she said, staring down at the ground.

Eventually, Pankaj continued onward, checking on the vaccinators going door to door. Then, when he was finished, we left. The road heading out of the village was a red dirt track and we rattled over it with our wheels in the ruts that the bullock carts had made.

"What will you do when polio is finally gone?" I asked Pankaj.

"Well, there is always measles," he said.

Jerome Groopman

The Biology of Hope

FROM *ACUMEN*

> *It's popularly believed that one's feelings can affect one's health, and that "positive thinking" can lead to a better outcome in fighting disease. Only recently, however, have researchers begun testing the "mind-body connection" in scientifically valid ways, and, as the physician and writer Jerome Groopman explains, their work has produced some surprising results.*

Not long ago, a woman whom I will call Julie Greene came to me for a second opinion. Julie was fifty-six years old and a school teacher. Four years earlier she had found a lump in her breast. A needle biopsy showed an invasive cancer; blood tests and X rays revealed that it had already spread to several ribs. Analysis of the cancer cells indicated that they lacked hormone receptors, and so would not be sensitive to treatment with estrogen blockers like tamoxifen; moreover, the tumor was negative for the amplification of the oncogene *HER2*, so therapy with trastuzumab, a recently developed antibody that targets this protein, was not appropriate. Julie received several courses of chemotherapy, and her cancer regressed. Then, three weeks before she consulted me, she felt a pain in her hip.

Tests showed the cancer was growing again, this time in the bones of her pelvis.

It was a crisp New England morning, the leaves turning their glorious colors, when Julie and I sat in the clinic to discuss her next steps. She had already seen two other cancer specialists besides her treating doctor. The therapeutic options I outlined for her were similar to those proposed by the other oncologists. "There are still a number of good drugs that may control the cancer—that is, cause the tumors in the pelvis to shrink and prevent any new ones from developing," I told her. Julie asked how long she could expect these backup drugs to work. I answered that the statistics I had were derived from groups, and that she should remember that people tend to seize on the average length of remission when reviewing data. But, I said, there is a bell curve, and some people are at the far end. Although her cancer was advanced, she might live for several more years. "I heard that already from my treating oncologist," she said. Then her voice grew heavy and tears began to well in her eyes. "One of the other oncologists told me that I had to have a positive attitude, and that the amount of anxiety I'm feeling is bad for my immune system. I know what that means." I paused for a moment and then asked her to elaborate. And what Julie said was what I had heard from many other patients. "How can I not worry and, at times, despair? Does that mean I'm helping to dig my own grave?"

Patients with maladies that have no ready cure are often burdened with information and misinformation about how feelings influence the outcome of their disorders. The past three decades have been marked by the ascent of alternative medicine, in which models of disease causation and treatment differ sharply from those of modern Western medicine. For much of those decades, each camp—the scientific establishment and the non-Western "healers"—has brusquely dismissed the other. One major point of contention has been the existence of a mind-body connection, the notion that emotions can be primary determinants of the genesis and eradication of disease. Recently, some Western physicians have adopted some of the tenets of alternative medicine: in a kind of hedge, they have proposed using the term *complementary medicine* to describe this integration. It was complementary medicine that Julie encountered when one of her consulting oncologists said her attitude could be deleterious to her immune system, that "negative" emotions like anxiety and despair would impair her body's defenses and give the malignancy free rein to grow and spread.

For many years, I wasn't very interested in how emotions might affect the outcome of illness. Of course, doctors of my generation used to pay attention to whether the vicissitudes of disease caused depression, and we prescribed appropriate medications for those patients who became clinically depressed. But any colleague who talked about a mind-body connection was met with a polite raised eyebrow or, less politely, a sneer.

This dismissive attitude was a reaction, in part, to the extreme claims made by prominent proponents of alternative medicine. A case in point is Norman Cousins. In 1976 this eminent editor published an account of his illness in the *New England Journal of Medicine*.[1] Mr. Cousins contended that he was afflicted with a collagen disorder and that none of the specialists at the University of California at Los Angeles School of Medicine could understand it. Lacking a precise diagnosis, they were unsure how to treat this serious autoimmune malady. Mr. Cousins was frustrated with their failure to arrive at a solution, so he repaired to a hotel room where he received intravenous infusions of ascorbic acid and watched comedies. Over the course of several weeks, Mr. Cousin's disease abated. He believed that the combination of laughter and vitamins was responsible for his remission.

Mr. Cousins's tale was widely retold and readily embraced by many desperate patients. It offered all the drama of a contrarian challenge to the scientific establishment. Moreover, it endorsed interventions that laypeople could pursue on their own. The subtext was empowerment; Mr. Cousins was telling patients that they could dispense with the canon tightly adhered to by the priesthood of physicians; any layman could take control of his body and rid it of pathology through emotion.

Yale professor of medicine Howard Spiro tried to track down the details of Mr. Cousins's apparently miraculous cure in his 1998 book, *The Power of Hope: A Doctor's Perspective*. Dr. Spiro is to be commended for his scientific detachment: he kept an open mind. But no details were ever forthcoming that could prove or disprove Mr. Cousins's story. After repeated requests for clarification, what Dr. Spiro finally received from Mr. Cousins was an autographed copy of his book. Absent any clinical information, Dr. Spiro speculated that Mr. Cousins may have had a self-limited illness, like a viral disorder, that spontaneously remitted. There was no proof whatsoever that laughter or vitamin C had anything to do with his restoration to health.

Alas, the lack of independent documentation in Mr. Cousins's case is typical of the literature of spontaneous remission. Stories about the powers of "positive thinking" usually lack rigorous data and fail to prove cause and effect. Feelings are depicted as magic wands that, if waved over the most refractory of maladies—cancer, lupus, AIDS—make them disappear. The darker side of Mr. Cousins's dogma is the concomitant belief that serious diseases can be caused by negative emotions. Clinicians like surgeon Bernie Siegel, author of the 1986 *Love, Medicine & Miracles*, contend that people who do not love themselves sufficiently, or who are unable to shrug off despair, anger, and resentment, unleash these feelings upon their own bodies and thereby trigger malignancies. There are no scientific studies to support such a hypothesis, but these views are still very common in popular culture. Susan Sontag in *Illness as Metaphor* has repudiated the notion of the "cancer character." A cancer sufferer herself, Ms. Sontag poignantly describes how this idea places the blame for her disease on her alleged defect as a person. The concept has another devastating consequence: by blaming the patient for her illness, the doctor implicitly divests himself of the onerous work of healing the patient. If the disease is not cured, it's because the patient was unable to muster the necessary good feeling to focus her immune system on the malignancy and chase it from her body.

It's easy to see why today's physicians, schooled in rigorous science, would flee from the mind-body hypothesis—at least as it is presented above. At best, it is wishful thinking; at worst, hurtful cant.

BUT BEYOND THE MORE EXTRAVAGANT CLAIMS, a very simple question presents itself. All physicians know there is a biology of fear, a biology of anger, and a biology of depression: might there also be a biology of hope?

I posed that question to Steven Hyman in 1999, then director of the National Institute of Mental Health in Bethesda, Maryland, and now provost at Harvard University. We sat together in his office, with its sweeping views of the NIMH campus. Dr. Hyman, a neuroscientist and a psychiatrist, readily endorsed the idea that positive emotions were an important field of study, but emphasized that the field was in its infancy. He said that one way to approach the biology of hope was by studying the placebo effect.

For many years, placebos were viewed as biologically inert—controls in the study of active intervention. Recently, scientists have come to understand that administering an inert substance can have a profound effect on neurochemistry and physiology if the patient doesn't know that it's inert. In fact, the placebo effect goes beyond the administration of the substance; a component of the effect on neurochemistry and physiology is the personal interaction between doctor and patient. Here belief and expectation are triggered by a doctor's words, gestures, and appearance, as well as by environmental cues like the framed diplomas on the wall.

Some of the most profound and reproducible changes wrought by a placebo can occur when a person is in pain. Neuroscientist Fabrizio Benedetti at the University of Turin, Italy, studies healthy volunteers who are subjected to a painful stimulus and given either an analgesic, like morphine, or a placebo, like saline. The volunteer does not know which he is receiving. The experiment goes something like this: the volunteer has a blood pressure cuff on his arm, and the cuff is inflated to exert a vise-like constriction. The volunteer describes, on a scale of one to ten, the degree of pain he feels at a specific pressure. Physiological changes associated with pain, like sweating, heightened pulse, and blood pressure, are measured. The cuff is relaxed, the researcher informs the volunteer that he will be given an analgesic, an opiate, and then the cuff will be inflated again. The drug is given, the cuff inflated. As expected, the opiate markedly reduces the participant's pain, and this is reflected in his vital signs. The sequence is repeated several times and then a sleight of hand occurs: instead of the medication, the subject receives a placebo. Yet despite the fact that the salt solution has no analgesic properties, the amount of pain the volunteer experiences is minimal. This is documented by the minimal changes in sweating, pulse, and blood pressure. But how can saline mimic a potent drug like morphine?

Dr. Benedetti has studied this in great depth, as have others, and they have found that the participants' expectations are sufficient to release chemicals in the brain that dampen the perception of pain. The chemicals include endorphins and enkephalins, both of which block the receptors on cells in pain circuits and thereby prevent the transmission of pain signals. Some recent work further suggests a link between dopamine-rich neural pathways that are part of reward-seeking behavior and the pathways that release endorphins and enkephalins.[2]

Researchers at the Karolinska Institute in Stockholm and Finland's Helsinki University Central Hospital studied the parts of the brain at work during the placebo response.[3] These neural reactions were compared to those triggered by a powerful and rapidly acting opiate. Networks of neurons recruited as the brain responds to a stimulus preferentially receive blood to provide the oxygen and nutrients those cells need for their increased workload. In the experiment, the shift in oxygen and nutrients was measured by positron-emission tomography, or PET, a scan that provides a picture of blood flow through the brain.

Nine people participated in the study. Pain was induced by applying uncomfortable heat for a little more than a minute on the back of the left hand. As a control, at other times during the course of the experiment, a probe at body temperature that was applied to the hand was sensed as neutral. The researchers, in authoritative scientists' garb and in the setting of a high-tech laboratory, informed each subject that two powerful analgesics would be used in the experiment, and that one of them was like morphine. But in truth, one of the two "drugs" that would be injected forty seconds before either the painful heat or the neutral probe, was saline.

There was an element of conditioning in the design of the Northern European study. During the week preceding the PET, each subject tested the hot stimulus, rating the intensity of pain on a uniform scale. This allowed the researchers to assess the degree of pain felt in response to the opiate or the saline placebo, as well as to gauge differences among subjects with respect to pain perception and stimulus response.

The PET showed that the analgesic activated a network of neurons in the cerebrum that prominently included the anterior cingulate cortex, important in cognitive reasoning, as well as in areas of the lower brain stem. There was significant overlap in neural pathways during placebo administration, again with increased activation of areas in the anterior cingulate cortex, as well as in parts of the frontal lobes just behind the eyes.

One of the other intriguing findings from the Swedish and Finnish study was that some subjects ("high-placebo respondents") had marked decreases in pain in response to the saline, while others ("low-placebo respondents") had less robust results. When given the opiate, high-placebo respondents' anterior cingulate cortices became active; the low-placebo respondents' did not. This suggests a relationship between how effectively drugs like opiates activate a discrete region of the brain and how strongly subjects respond to a placebo during pain.

We are all wired in different ways, the connections between our neurons sculpted in part by the genes we inherit from our parents, in part by past experiences, and in part by current environment. The way our brains process the thoughts and feelings involved in placebo administration, how we analyze a threat in a given context, what we expect of treatment, how much we desire relief, as well as other mental factors, can all strongly influence some of the most ingrained and fundamental aspects of behavior, like the perception of pain. But it also seems that thoughts and feelings influence our bodies' reactions to analgesics. It is a reasonable hypothesis, although one that must be subjected to the most rigorous scientific proof, that belief and expectation can influence the body's responses to other drugs, those that work in the brain, as well as those that work on other organs like the heart or lungs. It has been estimated that between one-third and one-half of the clinical effects of antidepressants can be ascribed to placebo response,[4] and this has generated much controversy and contention. The work from Northern Europe was a first step in what certainly will be a long journey of exploration, the goal of which is to explain how the mind exists in the brain, and how its intricate circuitry contributes to health and disease.

The differences between high-placebo and low-placebo respondents are interesting in another way too. Recent findings suggest that there are important genetic differences among individuals in dopamine-generating enzymes. If dopamine proves to be the key in the placebo-induced release of endorphins and enkephalins, then genetic heterogeneity in dopamine-generating enzymes will correlate with who has the most marked neurochemical changes attendant upon belief and expectation.

As a clinician, I think the above studies have profound implications. Pain is one of the most significant obstacles for patients facing difficult therapy and uncertain outcomes. Pain saps energy and wears down resilience. Pain often suggests to patients that they should not begin beneficial but arduous treatment, markedly reduces quality of life, and can cloud decision-making processes, rendering a person unable to choose among therapeutic options and begin on a path that offers the best chance for remission or cure. The clinical effects of placebos on pain are not trivial when seen in terms of a patient's larger experience of illness and the evolution of his clinical care.

Another recent advance in understanding how belief and expectation may alter neurochemistry and body physiology comes from research in Parkinson's disease. This is a neurological disorder in which dopamine is depleted in areas of the brain that are important for voluntary muscle control. Patients with Parkinson's develop rigid limbs and have increasing difficulty initiating and coordinating movement. Jon Stoessl, professor of neurology at the University of British Columbia, Vancouver, studied the placebo effect in patients with moderate Parkinson's disease. The motor function of patients was carefully described, and then some received a drug that releases dopamine in the brain; others received a placebo. The participants were told they could be getting active drug. As expected, with the drug that increased dopamine, patients' muscles became less rigid, and their voluntary motions more fluid. But more remarkably, the brains of patients who received the placebo showed the same amount of dopamine, as measured by PET, and similar objective improvements in spasticity and voluntary motor control. Again, it appeared that belief had sufficiently potent effects on pathways in the cerebrum to release neurotransmitters that modulated voluntary muscle function.

Other work on the placebo effect indicates that there can be important changes in other fundamental physiological functions. A placebo can mimic a bronchodilator and open up airways in patients with asthma: the chemicals that control airway narrowing are in the adrenaline family and can be released by autonomic nerves connected to the lungs. The neural pathways linking belief and expectation to these autonomic circuits have not yet been charted.

Belief and expectation can also contribute to outcome after myocardial infarction. Some, but not all, studies have indicated that patients who have hope, particularly those whose hope is based on religious faith, survive more often and stay in the hospital fewer days.

DESPITE THE MANY ANECDOTES about a mind-body connection, there is scant information demonstrating a clear-cut effect on the immune system and disease recovery. Richard Davidson, an experimental psychologist and the director of the W. M. Keck Laboratory for Functional Brain Imaging and Behavior at the University of Wisconsin, Madison, is one of the leaders in deciphering the components and consequences of hope. He has shown that people who are "resilient," able to cope with illness and loss and continue to function productively at home and at work, have a more normal pattern of cortisol release in response to stress than do others. Cortisol is a potent modulator of lymphocyte function, and homeostatic levels are important in both antibody production and the cellular immune response to viruses and fungi. In his studies of a cohort of women in Wisconsin who have been followed for decades since their high school graduation, Dr. Davidson has observed that the immune response to influenza vaccine is more robust among those who have a resilient character than those who do not.

Data from earlier studies further suggest that emotions associated with resilience affect immunity to infection with common viruses; a more rapid recovery is seen among hopeful people. But in an extensive review of the literature, I could not find convincing data on the effects of emotional state on the immune system of patients who have major diseases like AIDS or lupus. Furthermore, research that suggested that women with breast cancer who had an upbeat attitude and supportive friends and family had better outcomes has now been brought into question—including by the original investigators of the studies. Clearly, there are many variables in a clinical setting. Optimists with strong support may be more likely to pursue regular treatment and more likely to report complications during their therapy; therefore, they may receive more intensive and timely medical care. As much as many would like to believe that attitude is a powerful determinant of outcome in all cases, the data supporting this theory simply do not exist at the moment.

Does that mean that physicians should ignore emotions like hope that are associated with expectation? I believe not. Hope can overcome pain, and has observable physiological effects on respiration, motor control, and cardiovascular function. More crucial, hope is the foundation upon which cogent clinical decision making and a patient's resilience are built. Dr. Davidson says hope has a cognitive component and an affective component. The cognitive component is a clear-eyed view of difficulties, threats, and pitfalls. The hopeful patient is eminently reasonable. She sees a path through uncertain and possibly dangerous terrain to a better future. She takes into account all of the vicissitudes of a malady and all the limitations and side effects of a particular therapy. She makes the doctor her partner in the struggle to reach a remission.

The affective component, by definition, is suprarational. This is apparent in the metaphors we invoke to describe hope as something outside of ourselves that elevates and energizes us: we speak of finding hope; we say it has wings. The affective component of hope, Dr. Davidson posits, powerfully molds our capacity for resilience and sparks changes in our neurochemistry and physiology.

I told Julie (and her treating oncologist told her this, too) that the clinical course of breast cancer is highly variable, that in some women the disease can progress rapidly despite therapy, and in others it can be controlled for relatively long periods of time by treatment. We cannot predict what will happen to her; she might very well be in the latter group. We told her a tremendous amount of research was being done to test new treatments, both at academic laboratories and at pharmaceutical companies, and that some could fail but others might succeed. The clinical strategy was to try to stay ahead of her cancer, and that although there was currently no cure—total eradication of every malignant cell—periods of remission were possible. As to the mind-body connection, I told Julia that I knew of no data whatsoever that supported the notion that her natural feelings of anxiety or her moments of despair would accelerate the growth of her disease. As I write, Julie's tumors are shrinking through chemotherapy. It is too early to say how long her remission will last.

Restoring true hope to a patient is a physician's primary charge. I have often reflected on the famous phrase from the Talmud, the compendium of rabbinic thought and debate: "Where there is life, there is hope." In caring for the ill, we can invert that phrase: "Where there is hope, there is life."

1. Cousins, Norman, "Anatomy of an Illness (as Perceived by the Patient)." *New England Journal of Medicine* 295 (December 23, 1976), pp. 1458–63.

2. Jon-Kar Zubieta, "COMT *val158 met* Genotype Affects μ-opioid Neurotransmitter Responses to a Pain Stressor." *Science* 299 (February 21, 2003), pp. 1240–3.

3. Predrag Petrovic, "Placebo and Opioid Analgesia—Imaging a Shared Neuronal Network." *Science* 295 (February 7, 2002 [online]), pp. 1737–40.

4. Walter A. Brown, "The Placebo Effect." *Scientific American* 278 (January 1998), pp. 90–5.

Creepy-Crawly Care

FROM *SCIENCE NEWS*

> *For most people, just thinking about maggots is enough to induce queasiness. But as Ben Harder reports, doctors are rediscovering age-old therapies in which maggots are healing, rather than sickening, people.*

Pamela Mitchell is no stranger to modern medical care. Now fifty-two, the former waitress from Akron, Ohio, began getting regular insulin injections at the age of ten, after she was diagnosed with type 1 diabetes. Twenty years ago, she received a life-saving kidney donation from her brother. In 2001 Mitchell faced yet another harrowing health threat. An infected cut on her left foot had turned into a persistent, festering wound. Her doctors recommended that the foot be amputated.

That's when Mitchell began lobbying for a treatment that could have been concocted by the directors of *Fear Factor*. She pinned her hopes on a mass of writhing, munching maggots.

Hospital treatment of infected wounds with maggots—specifically, blowfly larvae—dates back more than seventy years. But use of these immature insects to remove dead and infected tissue, known as maggot-debridement therapy, long languished on the periphery of accepted practice. Improvements in surgical techniques and antibiotics displaced maggot therapy, as earlier advances had largely supplanted the medical use of leeches.

In 2004 both organisms received approval from the Food and Drug Administration to be marketed as medical devices. Neither had been strictly prohibited, but FDA's imprimatur, as well as some recent research findings, has given long-awaited credibility to the inclusion of live organisms in the medical armamentarium.

In the United States, maggots have been tested for use against several types of wounds and skin ulcers. Among these are diabetic ulcers, such as the one on Mitchell's foot, which nearly 15 percent of diabetics experience at some time. Roughly one in five of those ulcers eventually leads to limb amputation. In the United States, some sixty thousand limbs per year are lost to diabetic ulcers.

Another major problem that seems amenable to maggot therapy is bedsores, or pressure ulcers, which bedridden patients commonly develop. Sores increase both the length and cost of hospital stays. The most severe bedsores can lead to fatal complications.

In Europe and Israel, researchers have tested maggots in cases of burn injury, surgical wounds, traumatic injuries, and infections by flesh-eating bacteria. In all these conditions, removal of the dying tissue encourages healing.

While most U.S. medical institutions don't use maggot therapy, "more and more people recognize it as a viable option," says geriatric nurse practitioner Courtney Lyder of the University of Virginia in Charlottesville. He predicts that maggots will eventually rank second only to the scalpel in debridement, eclipsing tools he has analyzed in medical review articles, such as tissue-consuming enzymes. In some cases, maggot therapy may even be superior to surgery, Lyder says.

In 2001, as Mitchell's foot festered, scalpels and topical medications held sway in wound care. Her physicians cut away infected tissue—including the tips of two toes—and administered antibiotics, but those treatments didn't halt the spread of infection. Next, the surgeons told her that they would need to remove the entire foot.

"If I'd listened to them, I'd be in a wheelchair right now," Mitchell says. Having recently heard about maggot therapy on a television show, Mitchell urged her caregivers, including Akron-based dermatologist Eliot Mostow, to give the insects a chance.

Mostow, a faculty member of Northeast Ohio Universities College of Medicine, had used maggots therapeutically once before. He initially told Mitchell that her infection was too severe for maggot treatment, but when she remained adamant, he acquiesced.

He placed an order with pathologist Ronald A. Sherman of the University of California, Irvine, who raises, studies, and sells blowfly (*Phaenicia sericata*) maggots for medical purposes. Eggs of that species can be disinfected before they hatch into maggots. Just as important, the immature flies eat only dead and dying, or necrotic, tissue, which they break down by excreting digestive enzymes that don't harm healthy skin and bone.

Using a bandage, Mostow trapped the sterilized larvae in Mitchell's wound and let them feed for two days. After a series of treatments with fresh maggots over a period of weeks, Mitchell's wound was healing.

The maggots "really cleaned those ulcers out," says Mostow, who has since used maggots on several other patients. "Maggots do a phenomenal job of cleaning out dead tissue [in portions of a wound] where your curette or scalpel blade can't go."

But some patients find the procedure repulsive, Mostow says. When applied to the wound, day-old larvae are a millimeter or two long, but they grow to about a centimeter within a few days. The emerging larvae "are like Rice Krispy kernels, only oozing and falling all over each other," he says.

In Europe, where maggots are more widely used, some physicians place the larvae in permeable, plastic or nylon envelopes the size of tea bags. Enzymes and digested tissues can flow back and forth, while the growing maggots stay out of sight.

Mitchell still has her foot, and only a small scar to show for her ordeal. She now occasionally lectures doctors and nurses on maggot therapy and serves on the board of directors of the BioTherapeutics Education and Research Foundation. Sherman founded that organization last year to promote the study and use of maggots, leeches, and other living animals, including bees and pets, in medical care. This year, it began marketing "Medical Maggots."

Sherman is just the latest champion of maggots in medicine. Military surgeons of the Renaissance and in Napoleon's armies noted that wounds that became infested with maggots healed quickly and resulted in relatively few deaths or complications from infections. During the U.S. Civil War, at least one Confederate doctor applied maggots to soldiers' wounds. But the practice didn't leave the battlefield until after World War I.

In France in 1917 William Baer of Johns Hopkins University treated two gravely wounded American soldiers who, for a week, had lain unattended and unable to move in the contested turf between enemy lines. Maggots swarmed over the men's injuries, but beneath the larvae, the surgeon found surprisingly healthy tissue.

Baer was sufficiently impressed that, after returning to Baltimore, he used maggots to treat twenty-one nonmilitary patients over several years. After his results were published posthumously in 1931 maggots took hold in the medical community. By 1934, more than one thousand surgeons in North America and Europe were using maggot-debridement therapy for a variety of wounds.

From the 1940s onward, however, improved surgical techniques for debriding wounds and the advent of antibiotics contributed to the decline of maggot therapy, says Sherman. By the final quarter of the twentieth century, the practice was nearly forgotten.

Sherman, one of the few physicians who kept maggot therapy from disappearing in the United States, has spent more than twenty years researching the treatment of skin ulcers with blowfly larvae.

In 1990 Sherman began a study of the safety and effectiveness of using maggots in patients with persistent pressure ulcers. He followed 103 candidates for maggot therapy that were referred to him over the next five years.

Some patients did not receive maggot therapy because they, their family members or their primary physicians rejected the idea or because their wounds responded satisfactorily to conventional treatments. Others simultaneously received maggot therapy and conventional treatments, including surgery and topical gels.

Sherman collected data on ninety-two pressure ulcers—some as large as a business card—in sixty-seven patients. Of these wounds, forty-nine were treated with conventional means alone and forty-three were treated with maggots either alone or with conventional therapy.

During the first month of treatment, 79 percent of the wounds treated with maggots decreased in surface area, and the average area of dead or dying tissue fell by a third. In wounds that didn't receive the maggots' ministrations, only 44 percent shrank within the first month and there was no overall change in necrotic area. Within five months, 39 percent of the maggot-debrided wounds had healed completely, whereas only 21 percent of the others did so. Sherman reported these findings in 2002.

Sherman performed a separate evaluation of maggot therapy for treating patients with diabetic foot ulcers. He examined twenty such wounds in eighteen patients, whom he treated between 1990 and 1995. These wounds received maggot therapy, conventional surgical and topical procedures, or both approaches.

His results, reported in the February 2003 *Diabetes Care*, resemble his pressure-ulcer data. During the first two weeks of maggot therapy, sometimes combined with conventional treatment, the average percentage of each wound covered by necrotic tissue fell markedly, and 36 percent of those wounds ultimately closed completely. Conventional therapy had no significant effect on the necrotic area during the first two weeks, and only 21 percent of the conventionally treated wounds ultimately healed fully.

Maggot therapy may carry yet another advantage. Among ten deep wounds that were cleaned by maggots before surgical repair, there were no postoperative infections. But there were six infections among nineteen similar wounds that were not exposed to maggots, Sherman and Kathleen Shimoda of Veterans Affairs Long Beach Healthcare System in California report in the October 1, 2004, *Clinical Infectious Diseases*.

While few other researchers in the United States have reported data on maggot therapy, several teams in Europe and Israel have been studying the use of the larvae during the past decade. A few studies have reported that some patients with painful skin wounds experience an intensification of pain during maggot treatment. However, most of the studies suggest that maggots are at least as effective, and possibly less expensive, than conventional therapy for infected skin wounds.

A decade ago, Sherman estimates, at most twenty vials of maggots were used in North America each year, excluding those he used himself. Last year, he shipped about one thousand vials to U.S. doctors. Maggots, Sherman says, "have found their way back into the hearts and minds of many clinicians, and into the wounds of many patients."

Jennifer Ackerman

Cranes

FROM *NATIONAL GEOGRAPHIC*

Graceful, easily recognizable, celebrated for thousands of years in many cultures, the crane is a popular and beloved bird. Nature writer Jennifer Ackerman follows a group of "craniacs" who are trying to bring these remarkable species back from the edge of extinction.

From a blind overlooking the wetlands of central Wisconsin, I can see a long-legged bird in the distance, a stroke of white curled at the top, like a bright question mark against the emerald green grasses. Then up pops another from the screen of reeds. The birds are yearlings, five feet tall, with snow-white plumage and elegant black wing tips that spread like fingers when they fly. They're quiet now, but from the long trachea coiled in their breastbones may come a wild, singing whoop, harsh and thrilling, that gives their tribe its name.

This would be a primordial scene—big sky, undulations of tall marsh grasses, wild whooping cranes—were it not for a penned area nearby, where several whooper chicks, well camouflaged in tawny feathers, forage in the shallows. In a whisper, crane biologist Richard

Urbanek explains that these chicks have been raised in captivity but have never heard a human voice nor seen a human form, except in crane costume. As part of an experimental program to reintroduce a wild migratory population of whooping cranes to the eastern half of North America, these chicks have been fed and tended by crane-costumed people for two months. Now, before they are released to the wild, they are being taught the habits of their ancestors with modern techniques pioneered by Operation Migration, an organization devoted to helping endangered birds learn their traditional migratory routes. Near the pen is a long stretch of open grass, a runway, where the chicks are learning to fly behind an ultralight plane flown by a pilot in crane costume who will guide them from this refuge twelve hundred miles south across seven states to wintering grounds in Florida.

Two cohorts have already made such trips—and returned on their own, the first whooping cranes in perhaps more than a century to fly freely over the eastern United States. After three years of ultralight-led migrations, the new eastern migratory population numbers thirty-six birds, including the yearlings and the chicks. The success of this effort is leading the way for a more ambitious project half a world away in the northern reaches of Russia. In the fall of 2005 an international team plans to lead a flock of young captive-bred Siberian cranes along part of their traditional migratory route, from Russia to Iran, to restore the birds' knowledge of the ancient flyway—not with ultra-lights but with hang gliders that will soar a difficult path extending more than three thousand miles over four different countries.

These human-guided migratory flights are among the most recent acts of vigorous intervention to rescue from extinction a singular creature—what conservationist Aldo Leopold called "no mere bird" but "wildness incarnate." For thousands of years cranes have been honored for their beauty, their ancient ancestry, impressive size and flight. In Africa and Europe their image appears in prehistoric art. They figure on Egyptian tombs, in Russian songs, in the totems and clans of Native Americans, in Australian dances, and Greek and Roman myths. In many parts of Asia cranes are held sacred as symbols of happiness, good luck, long life, peace. After the dropping of the bomb that people said was brighter than a thousand suns, a young girl stricken with radiation sickness set out to fold a thousand paper cranes in the hopes that she would recover. She died before reaching her goal, but other children pursued the task, and now the stone monuments of Peace Memorial Park in Hiroshima are ornamented with millions of the tiny folded cranes.

The esteem in which these birds are held has not spared them destruction. Cranes are among the most endangered families of birds, having been hunted, persecuted, chivied out of their last havens by human pressures. Nine of the planet's fifteen species are threatened with extinction. In East Asia loss of wetlands threatens the red-crowned, the hooded, and the white-naped cranes. The graceful blue crane, the national bird of South Africa, has suffered from predation by wild dogs and the cultivation of tree plantations, which has eliminated great stretches of its unique grassland habitat.

Whoopers, the rarest of cranes, were extirpated from much of their range in North America in the nineteenth century by hunting, egg collecting, and habitat destruction as settlers drained wetlands and plowed prairie for farming. By the early 1940s only twenty-one birds remained. The extreme plight of the whooping crane alerted many people to the high price we may pay for harming the natural world—and the need for extraordinary efforts to recover what is almost lost. With the help of habitat protection, hunting restrictions, and captive-breeding programs begun in the 1960s, the remnant population of whooping cranes began to grow. It now numbers three hundred birds in the wild and more than one hundred in captivity—not anywhere near its original abundance, but a big step on the road to recovery.

One man who has led the strenuous work to save the whooping crane is George Archibald, cofounder of the International Crane Foundation (ICF) in Baraboo, Wisconsin. Now Archibald has a new dream. In a patch of prairie spangled with blooms of butterfly milkweed and *Silphium* not far from the headquarters of the ICF,

Archibald sits in the same dark shack where Aldo Leopold wrote his classic essays on conservation. A self-described "craniac," Archibald is considered the world's foremost expert on cranes. He is an unassuming man, committed to rescuing cranes from the abyss of extinction, and forever optimistic about his chances for success. In the past quarter century he has helped to launch crane conservation programs in Japan, China, South Korea, India, Iran, South Africa, Australia, and Russia. He and his team at ICF have created a species bank of captive cranes to guard against extinction. He has carried crane eggs tens of thousands of miles in plywood boxes and delivered them safely to captive-breeding facilities. He has danced with cranes, sung with them, devoted his life to saving them and their habitat.

Why? "Cranes are ambassadors of the ecosystems in which they live and also of international goodwill," he says, "two things we need to conserve. Because these birds require pristine habitat, they act as umbrella species; if you save them, you are also saving the wetland and grassland ecosystems on which they depend. And because their migratory routes don't heed political boundaries, any effort to protect them requires the participation of diverse people in different countries. So they act as vehicles for cooperation between nations that are often politically polarized." He pauses. "Also, I love them."

Archibald's latest goal is to restore a population of migratory Siberian cranes to Central Asia. Sometimes called snow wreaths, the magnificent Siberians are the most highly specialized of cranes, depending exclusively on bogs, marshes, and other wetlands for nesting, feeding, and roosting. They are also the most critically endangered, their numbers dwindling from loss of habitat and from hunting during migrations. The Siberians have traditionally migrated more than three thousand miles from the high tundra of Siberia across eleven countries to wintering grounds in China, India, and Iran. The success of the migration depends on the welfare of chains of wetlands across the continent, which serve as stopover points for the birds.

Archibald and an international team are working to secure legislation safeguarding these areas and upgrading their protection. The effort takes patience, as many of these nations are dealing with difficult political situations, limited resources, and skeptical leaders.

In the summer of 2005 the team plans to raise a dozen Siberian chicks in captivity with crane-costumed parents and train them to fly behind hang gliders flown by crane-costumed pilots. Then, sometime that autumn, three hang gliders will launch from Uvat, Russia, soar over western Kazakhstan, stopping at the Volga River's Astrakhan Nature Reserve, then wing south over Azerbaijan, along the western shore of the Caspian Sea, to the flooded fields at Fereydun Kenar and Esbaran, the cranes' wintering grounds in Iran. The birds will follow.

The international effort to find these birds safe passage in such politically troubled areas is considered crazy by some and brilliant by others, a way of drawing global attention to the cranes and their world—minnow, bulrush, cattail, river shallow, aura of ancient flyways that unite regions and ignore borders.

Edward Hoagland

Small Silences

FROM *HARPER'S*

*For the naturalist and essayist Edward Hoagland, nature is not sim-
ply what's outdoors, but something that is intertwined with our own
human nature. In this evocative memoir of growing up with nature,
he confesses his beautiful love affair with the natural world, and
wonders whether our humanity will survive nature's encroachment.*

Wandering to the edge of Dr. Green's woods, next door to
our new house, at the age of eight, I found a little brook
running—my first, because we had just moved from the
city to the country. The floating twigs and leaves, the ripples, and yet
the water's mirroring qualities, and the tug on my fingers or feet when
I dipped them in, plus the temperature, so remarkably different from
the air's, fascinated me. We, like the other neighbors, had some brush
and untrimmed trees in back of our lawn, so the brook picked up
interesting flotsam before entering Dr. Green's pines, where except for
scattered boulders or stones the ground was less varied, all needly.
The sounds the stream made, thocking and ticking, bubbling and
trickling, were equally beguiling, however, and, like the wind-nudged
boughs twisting overhead, never precisely the same.

This was in Connecticut during World War II, so we kept a dozen brown hens for their eggs, and I watched their pecking order develop, and other habits, endlessly. But the brook was a near second for curiosity, and because it was undomesticated I recognized in it a wilder power. I felt like part of the flock—feared for them at nightfall, when they would duck inside the coop through an entrance I never forgot to close, and rejoiced with them when a New Hampshire Red rooster was acquired to trumpet their accomplishments. But the woods were an adventure, more mysteriously reverberant. By nine I was probably daring enough to follow the stream (knee-high, so nobody was afraid I might drown in it) through the forest far enough to catch sight of Dr. Green's pond and approach its spongy inlet. I must have heard about its existence, but no one had taken me there—my father golfed, my mother gardened. The distance is now elastic in my mind's eye, but no roads or houses intervened and my discovery caused no alarm. My parents, retracing the path, notified Dr. Green, a retired and retiring widower whose given name I never learned. They owned two acres; he, I would guess in retrospect, at least fifteen. Then there was Miss Walker's estate, in back of both properties, and lending them a bit more wildlife, such as the foxes that threatened my chickens. She was a fiftyish spinster, with servants, who may have had about three times as much land. My parents had never met her, but she caught me trespassing as a "nature boy" once or twice—climbing the spruces in her overgrown fields—and summoned me inside her manorial stone house (ours and Dr. Green's were white clapboard) and gave me and the friend who was with me a proper glass of apple juice with chocolate-chip cookies. A real-estate agent who wanted to subdivide the place surprised me too and was less friendly, afraid that I might start a wildfire, but I would sooner have harmed myself, and I think Miss Walker surmised as much and knew you couldn't stop a boy from crossing stone walls and wire fences.

The pond was the great revelation, after, first, the stream and before I could climb a sixty-foot Norway spruce and swing with the wind. Amber, black, and silver, with moss on one bank and cattails on another, it had frogs plopping—leopard and green frogs and bullfrogs with large eardrums who said jugarum. I saw crayfish and a ribbon snake, yellow-striped, and a muskrat swimming in a moving V. Greenish pollywogs and salamanders. The pines, in retrospect, were red pines, and red squirrels of course were chattering in them from all angles, much more vociferous than the gray squirrels that nested in the maple shade trees in front of our house. Also my dog, Flash, an English setter and constant companion, who barked at them, sniffed out an opossum, longhaired, long nosed, pouched like a miniature kangaroo, who promptly swooned and played dead when I picked it up by the tail, just as it was supposed to do. The pines, with their thousand jewely shards of light as you looked up on a sunny day, didn't like wetting their feet and gave way to white birches whose curling strips of bark you could write on not just with a pencil but with your fingernail if you had to, and dark hemlocks, droopy-branched and unbelievably tall and somber, in the soggy ground, while the pond itself might be as bright as a lens of glass, with tree crowns, mackerel clouds, and blue sky reflected on its surface.

I was a good student and not as friendless or solitary as this may sound. My trouble was a bad stutter that made Flash's companionship and the flock of chickens and communing out-of-doors important to me. But I generally had an intimate friend of my own age whom I could lead to these precious places—Miss Walker's lordly spruces, like the beanstalk that Jack climbed, or Dr. Green's miraculous, jam-packed pond—discovering, however, to my astonishment, that they didn't matter as much to him. He would prefer listening to Mel Allen broadcast the Yankee games with me, gossiping about our class, inventing piratical schemes. Ice-skating, yes, we shared, or throwing rocks into the water to watch the thumping splash, but when we were past the volcanic stage, why go? Dr. Green was a cranky, softhearted soul almost too old to walk to his pond by the time I discovered it, but he made the effort to inspect both me and it when he heard from his gardener that I was there, telling me not to treat the frogs inhumanely or to fall through the ice. Brooks, the silent Irishman whom he had long employed, had no interest in woods or ponds, so I seldom encountered him, and he was easy to avoid, sitting smoking in his chilly greenhouse—just nodded at me as an authorized visitor if we did chance to meet.

I saw a mink trap an eel, and a musk turtle diving out of reach on the bottom, and tree frogs in the bushes, and "pumpkinseeds," or sunfish; learned how to fish with a worm and a pole. A blue heron bigger than me would sometimes flap over from the larger ponds (Dr. Green's, though huge to a ten-year-old, may have been scarcely an acre in size) after amphibians. Or on my repeated expeditions I might see a kingfisher diving for shiners. Nobody could spot me from the road as I circled the water, looking at the strange jointed plants called horsetails, or ferns at the margins. The ordeal of stuttering at school seemed distant indeed, and I was learning how to swim well enough at my parents' country club not to be a danger to myself if I fell in. I was cautious anyway, as stutterers must be if they are to survive— bicycled home after swimming lessons, before I'd have to try to talk to people on the patio. I could talk to Flash with absolute abandon, and loved Mel Allen, and Tommy Henrich, "Old Reliable," the Yankees' right fielder, and watched the New Hampshire Reds' social goings-on like a budding ethnologist, though I tended to downplay my various excitements in the house lest they be restricted or used against me. It was not a silly instinct because my parents did soon tell me I was reading too much, and by prep school were telling my favorite teachers that I was too intrigued by nature and writing; that these were dodges due to my handicap and might derail a more respectable career in law or medicine—angering the teachers who nurtured me.

I WAS AN ONLY CHILD UNTIL, at ten, we adopted a three-month-old baby girl named Mary Elizabeth, from Chicago. So this event may have been another force pushing me into the woods. Within a few years she and I became close, though in the beginning I used to proclaim that she wasn't my "real sister" in order to watch my mother's distress. And from the pond, after school, I would often go uphill, not back along the stream, to Dr. Green's house, skirting both Brooks and, in particular, the doctor himself. That is, he was always indoors, but I would take care not to be heard in his bedroom or sitting room upstairs, and knock softly instead at the kitchen door so that Hope alone heard me. She, as cook, like Brooks, had worked for Dr. Green for many years but had not established the same fellow feeling. She was brown-skinned and tacitly more intelligent and ironic than Brooks was, and took good care of her employer without receiving the hearty although formal daily greetings Brooks did. He was white and had a family in town to repair to, which also made him more of an equal, perhaps, than a large-waisted, large-breasted, light-colored Negress in a pink uniform, far from wherever her grown-up children were located, or other friends. Being small and mute, I never inquired of Hope where she had originated. I was grateful for her kindness in inviting me in, to sit on a white chair, with the kitchen clock ticking and the scents of baking. Like the pond, it soon became a dependable sanctuary where nobody asked me to speak. We simply sat quietly, she with her hands in her ample lap, while Flash and the doctor's cocker spaniel lay down together on the back porch. Sometimes I forgot my handicap, but if we talked for a minute it was in low tones so that the lonely doctor wouldn't tromp downstairs and interrupt us, asking angrily why I hadn't come to see him. To a little boy he appeared formidably crabby and diagnostic—naturally wanting to hear, analyze, and cure my stutter.

Jimmy Dunn, in the house on the other side of ours, was a good playmate for cards and chess and imaginary games; and Tommy Hunt, at whose house the school bus stopped, was a likable guy—he eventually became a minister—more mature than us, who worked on a jalopy about as much as I went to the pond and had a crush on the girlfriend of our football quarterback, who became an airline pilot after high school and college, when she married him. By eighth grade I had a silent crush on her too; she was our most down-to-earth, approachable blond. Though nobody picked on me, some of the other boys collected at Tommy's after the bus dropped us off for BB-gun fights or to drive that hoodless jalopy around and around Tommy's parents' vacant field or to squeeze each other's balls in the bedroom in a manhood test or to tie each other up or masturbate their dogs, including mine if they could catch him (to my cowardly shame). I sometimes hunted rats with Flash in our relic barn with my own BB gun, after I had ceased to need to seek out Hope so much. But the wonder isn't why these kids didn't beat me up—even the Kane boys, sons of a drunken "black Irish" gardener, gap-toothed, tricky and sly, who threw stones at cars and went to the public school, not ours, did not, though I was certainly warier when crossing their employer's meadow—as much as why nobody ever has. Not when I honeycombed Boston on foot for years and New York for decades in late-night walks, or in five trips to Africa and nine to Alaska, and so on. I don't need to belabor the point, but neither cowardice nor caution alone explains it. A naturalist's and a stutterer's intuition maybe more so—I've swung around and looked into an approaching mugger's eyes occasionally, which caused him to sheer off—and a berserk streak that I have when sufficiently angered that perhaps sets other people back on their heels long enough for a bad situation to defuse. Stutterers learn to distinguish genuine danger from the ersatz, and also to manipulate their anger for the fluency that a shot of adrenalin will momentarily bestow. And they may develop a well of empathy that, again, deflects the rage of sufferers looking for someone to attack, and learn to distinguish the fulcrum of power in any group. While bullies, for example, will persecute the closest target, real predators look into the middle distance—as I learned in my late teens when taking care of lions and tigers in the Ringling Bros. and Barnum & Bailey Circus. So you avoid a bully, but stand in closer proximity to a predator and join him in gazing out.

Girls were another puzzle. One of my sister's baby-sitters would coax me to lie on the sofa with my head in her lap, before I was ready to, so that she could practice what she wanted to do when she was with her boyfriend. Or she would ask me to reach for something on a high shelf, then press her breasts against my back, as if to help. At a dancing class another girl taught me to cut in on her when she was burdened with a partner she didn't like, but after I took this to mean she was partial to me she got mad because I cut in on her when she was fox-trotting with a boy whom she did like. Dr. Green died while I was away at boarding school, so I can't recall whether Hope retired somewhere down South, being of an age to, or sought other employment. But she was authentically welcoming to me, and suitably erotic to a prepubescent too, sitting

catty-corner to me across from the big old-fashioned stove and oven—
with invariably a roast in it for the solitary old man upstairs—and her
hands loosely clasped, a gentle expression, big hips, lax bosom, and her
uniform collar unbuttoned. I was never interrupting anything; she'd
have the clock to listen to, the radio on low, a newspaper the doctor had
finished with folded on the counter. Live-in maids in our neighborhood, black or not, had no cars to visit one another and would not have
been permitted by the police to walk the roads. I took care not to trigger my stammer by telling her I had just seen watercress, water striders,
eelgrass, mudpuppies, duckweed, pickerelweed, horned pout, and
water with trembling algae, wavy larvae, and waterlogged trees like
slumbering alligators three feet down. The clouds had piled up like
smoke signals over the pond and the pines, and I had nibbled rock tripe
and touch-me-nots. One winter I'd accidentally angered a friend by
grabbing at him as I fell through the ice, pulling him into an icy bath.
Our families thought, oh, dirty water, but I knew that, no, it was where
life lived, and part of my heart. By twenty I would be climbing unob-
served into a mountain lion's cage, but already I trusted my faith in
nature and was biking to larger ponds, then every week or two basking
in my secret refuge at Hope's kitchen. I didn't eat there or stay as long
as an hour, just dropped in to decompress, resting.

Hope was not the first colored person I'd known. (The term
"black" then was an epithet of insult, equivalent to "nigra.") When my
mother brought my baby sister home on the train from the Chicago
adoption agency, she had hired a woman from the South named
Arizona, much younger than Hope, vigorous, boisterous, taller,
darker, and less acclimated to the behavior expected of servants in an
upscale Wasp suburb up North. She was a blithe spirit, as I remember
her, assertive, gleeful, expansive, loud and goofy with me when, to
tease her, I'd pull on the bow of her apron strings—which, though I
was about ten or eleven, quickly alarmed my Missouri-born father as,
I suppose, proto-miscegenation. In a few years he would begin boy-
cotting the Metropolitan Opera for permitting the great black con-
tralto Marian Anderson to sing on its stage. Arizona had big buttocks
under the thin fabric of her uniform that made the butterfly of the

bow doubly tempting, and he perceived a sexual element in our gig-gles. My mother, not being from a former slave state, was more star-tled by Arizona's guileless tales of growing up barefoot in a hovel, and chewing her own babies' meals for them when she couldn't afford to buy prepared food. It sounded unsanitary, barbarian, "African," like her strangely untrammeled name (evocative to me now of wanting to "light out for the Territories") and her maid's-day-off visits every Thursday to Harlem by taxi and train. Goodness knows what diseases she might bring back to her attic room and our dishes and pantry. My mother rang a tinkly bell at the dining-room table, rather than possi-bly interrupt a polite conversation with guests, to call Arizona to clear each course off the table, but Arizona didn't seem to fathom the gen-tility of the ritual; was likely to holler out helpfully to ask what was wanted. So they got rid of her, my father believing afterward that it reduced a property's value, at least in Connecticut, if a colored person had ever lived in it, even in a servant's capacity.

I learned from the episode not to betray to a third party affection for anybody who might get fired because of it, or to divulge any passion that might thereafter be denied me. "I'm going to the pond," I'd say casually to my mother; then dodge carefully past the stolid, deracinated Brooks (like a tug-on-the-forelock-when-the-gentry-go-past footman), toward the trillium and columbine, the toadstools and fairy-ring mushrooms, the nematodes and myriapods, the blueberries or blackberries, near the opaque yet shiny stretch of hidden water, deep here, shallow there, with the wind ruffling the surface to conceal such factual matters, and cold at its inlet but warm where it fed into a creek that ran to the Silvermine River and finally the ocean. Getting hold of a live-trap, I caught a couple of weasels that screamed at me through the mesh until I released them, and a burgler-masked coon, and the inevitable beautifully white-caped skunk, who didn't let me have it when I let him go.

The plopping raindrops, wobbly riffles, crosscurrent zephyrs, the penny-sized and penny-colored springs that replenished the margins of the pond from underground, pluming hazel-colored, endlessly rising-and-falling individual grains of sand, irislike around the black pupil of the actual hole, lent variance to the velvet water, near dusk, or bright mornings, when it shone in the mini-forest like a circlet of steel. During a thunderstorm it seethed, fingered madly. Then when the clouds cleared off, in the batty moonlight, the shadows seemed crafted differently than might be cast by any sun. I was delighted watching bats flutter after hatching mosquitoes in the wetland that bulwarked the pond, and thrilled when the pair of barn owls that nested in an abandoned water tower on the hill would shriek as they skimmed across Miss Walker's second growth. It was more frightening to be alone upstairs in my own house at night than to tiptoe about the woods. Nor was it as scary peering at a copperhead on a ledge one noontime that a schoolmate whose mother was a birdwatcher took me to see. And I loved the whirligig beetles and water boatmen, the damselflies and fireflies, the sticklebacks and freshwater snails. My favorite turtles were the wood turtles, *Clemmys insculpta,* seven or eight inches long, that have almost disappeared from New England now, with their sculptured carapaces, like Cellini's metalwork, and salmon-red legs, which I would watch breed in the stream in the spring, but that roamed the fields until hibernation time, when they'd return again to the streambed's leaves and mud.

Snappers lurked in the muck of the pond year-round, platter-sized, but didn't bite if you left them there, even if you happened to step quite close or purposely touched their serrated tails: only if you picked up a female on her big day in June when she left the water to lay her eggs. The bottom, when you waded, was painted with fallen leaves and so varied you'd stand with one foot sucked ankle-deep and the other supported on hard feldspar sand—little fish angrily nipping at you because you had begun to infringe on their nesting terrain— and one warm but one chilled, and a branch knifing out of the water alongside you much like a fin. Wood frogs and peepers were to be found in the bushes, and a water snake tasseled in lovely russets and tans somewhere down the bank. A green frog or bullfrog couldn't survive the snake's visits simply by holding its breath underwater; it had to swim in zigzags or leap fast. But speed was not a defense when the great blue heron, gangly legged, slow flapping, maneuvered down from the sky. It had a spear beak and watched as sternly as a sentinel, once it had landed, for the first frog buried in mud that needed some air. Similarly, a mamma duck might be clever at concealing her ducklings in the reeds from me—but not from the mamma snapping turtle, who grabbed them one by one through the late spring before laying her own eggs. Even a seagull dropped by from Long Island Sound to forage on crappies or whatnot, and Mrs. Morris, my sixth-grade biology teacher, told me that my eels, too, had migrated in from the sea. The Canada geese barked like beagles, going north or south.

On sunny days a certain woodchuck liked to clamber up onto the leaning bough of an apple tree next to the stream and straddle it as comfortably as if she thought she were not just a matriarch but an arboreal creature. I'd face the ethical dilemma of whether to notify Flash and provide him some fun at the expense of panicking her miserably. An old apple tree's outreach, like that one's, carries an idiosyncratic eloquence because season by season the weight of its fruit has twisted each individual limb. This generosity speaks, whereas a white spruce's symmetry is more visually generous, and climbing high to rock with the wind was to plumb a power no truck tire roped like a pendulum to a maple or an oak tree could approach. I'd lie on my back on a patch of moss watching a swaying poplar's branches interlace with another's, and the tremulous leaves vibrate, and the clouds forgather to parade zoologically overhead, and felt linked to the whole matrix, as you either do or you don't through the rest of your life. And childhood—nine or ten, I think—is when this best happens. It's when you develop a capacity for quiet, a confidence in your solitude, your rapport with a Nature both animate and not so much so: what winged things possibly feel, the blessing of water, the rhythm of weather, and what might bite you and what will not. In the circus a tapir, a tiger, a mandrill, a rhino, but building really upon the calm that Dr. Green's modest woods and pond, forty-five miles out of New York City, had bestowed on me.

NATURE INDOORS—that plump bobby-soxer stroking my hair while holding my head in her lap to practice up for her boyfriend—

made me more jittery, but I was not really somebody who "liked animals more than people," as the cliché goes. Animals didn't sometimes smile sardonically or in wonderment when I stuttered and avert their eyes, turn their backs, but I had close friends and was enjoying my sister's presence now too. I loved her, and even found she was deflecting about the right amount of attention away from me in family politics (but not too much). I didn't go swimming in crashing surf or lightning flashes or climb cliffs with ropes and pitons or kayak in whitewater rapids or spelunk claustrophobically. I wasn't trying to conquer nature or prove my testosterone. But nature as simply night or a height or a lonely menagerie animal or a small limestone cave to crawl down in or the lip of the crest of an unpretentious mountain to hunker on for an hour felt just right, and often as if my throttled mouth and bottled-up emotions had engendered a sort of telepathy in me. Not of course to warn of inanimate events, like a flash flood or a rock slide, but the bear around the bend or the desired milk snake in the woodshed. My sixth sense was unstoppered.

I never totaled a car (machines may not have interested me enough) or broke my bones, and had an upbeat view of life, experiencing the kindness of many strangers when I hitchhiked, for instance. I speculated as to what the anthropological purpose could be of the brimming, broad-gauge affection people like me felt when watching a wriggling tadpole or clouds wreathing a massif—sights that have no reproductive or nutritional aspect. Call it "biophilia" or agape; it wasn't in response to a hunter's blunt hunger, or kinship-protective, or sexual in some way. Was it a religious wellspring, then? Silence and solitude are fertile if the aptitude is there, and love in its wider applications is also, I think, an aptitude, like the capacity for romantic love, indeed—stilling for a few minutes the chatterbox in us. That massif wreathed in clouds, or the modest pond that has been left in peace to breed its toads, is not a godhead. Like sparks flung out, each perhaps is evidence instead (as are our empathy and exuberance), but not a locus. And yet a link seems to need to take hold somewhere around nine, ten, or eleven—about Mowgli's age, in Kipling—between the onset of one's ability to marinate in the spices of solitude, in other words, and puberty, when the emphasis will shift to contact sports, or dress and other sexual ploys and fantasies or calculations.

But nine was fine; and when you came to feel at home in Connecticut's woods, New Hampshire's were not a large step up the ladder, or Wyoming's expansive mountains after that, then California's by twenty, building toward British Columbia's and Alaska's, Africa's and India's, in the course of the future. The sea was different, however. I admired it from the beach or a steamship but never acquired the nonchalance required for solo sailing; was afraid of drowning. On the other hand, having been born in New York City and then returned to live there as an adult, I loved metropolises and saw no conflict between exulting in their magnetism and in wild places. Human nature is interstitial with nature and not to be shunned by a naturalist. This accidental ambidexterity enriched my traveling because I enjoyed landing and staying awhile in London on the way to Africa, or exploring Bombay and Calcutta en route to Coimbatore or Dibrugarh. Didn't just want to hurry on to a tribal or wildlife wilderness area without first poking around in these great cities, which I rejoiced in as much. Although there are now far too many people for nature to digest, we are all going to go down together, I believe. We are part and parcel of it, and as it sickens so will we.

In the meantime, joy is joy: the blue and yellow stripes of a perfect day, with green effusive trees and the dramatic shapes of the streaming clouds. Our moods can be altered simply by sunlight, and I found that having cared for primates, giraffes, and big cats in the circus made it easier to meander almost anywhere. Few people were scarier than a tiger, or lovelier than a striding giraffe, or more poignant than our brethren, the chimps and orangutans, and you can often disarm an adversary if you recognize the poignancy in him. Nevertheless, I preferred to step off the road, when I was walking in the woods at night and saw headlights approaching. Better to take one's chances with any creature that might conceivably be lurking there than with the potential aberrations of the drive-by human being behind the wheel. It may seem contradictory that for reverence and revelation one needs a balance. You can be staggered by the feast of sensations out-of-doors, but not staggering. Your pins ought to be under you and your eyes focused. As in music, where beauty lodges not in one note but in combining many, your pleasure surges from the counterpoint of saplings and windthrow, or the moon and snow. Both are pale and cold, yet mysteriously scrimshawed—the moon by craters, mountains, and lava flows, the snow by swaying withes or maybe a buck's feet and antler tines. Although like snow, the moon will disappear predictably and reappear when it's supposed to, moonlight is an elixir with mystical reverberations that we (an pine and yet grin over, even when "empty-armed." It's off-the-loop, a private swatch of time, unaccountable to anybody else if we have paused to gaze upward, and not burdened with the responsibility of naming birdcalls, identifying flowers, or the other complications of the hobby of nature study. One just admires a sickle moon, half-moon, full moon that, weightless and yet punctual, rises, hovering. Sometimes it may seem almost as if underwater, the way its dimensions and yellow-ruddy coloring appear to change—butter, or russet, or polar. The Hungry Moon, Harvest Moon, Hunter Moon, are each emotional, and expertise about their candlepower or mileage from the earth a bit extraneous. Although our own cycles are no longer tied to whether they are waning or gibbous, we feel a vestigial tropism. This is our moon. It's full, we'll murmur; or It's a crescent, or like a cradle lying partly tipped. And a new moon is no moon.

Twilight, the stalking hour, itself can energize us to go out and employ that natural itch to put our best foot forward and "socialize." The collared neck, the twitching calf, and tumid penis will respond to daylight's variations or the moonrise, as we gulp raw oysters and crunch soft-shelled crabs that still possess that caught quality, not like precooked pig or processed cow. If we've lost the sense of astrological spell and navigational exigency that the stars' constellations used to hold, we at least present fragrant bouquets and suck the legs of briny lobsters like savages on important occasions. The stunning galaxies have been diminished to blackboard equations that physicists compute, and our dulled eyes, when we glance up, instead of seeing cryptic patterns and metaphors, settle rather cursorily for the Moon.

Water does retain a good deal more of its ancient power to please or panic us. Bouncing downhill in a rocky bed, shouldering into any indentation, and then nurturing fish, mirroring a spectrum of colors, or bulking into waves that hit the spindrift beach at the inducement of the wind, it's the most protean of life's building blocks, the womb of the world. "My god, there's the river!" we will say, in pure delight at the big waterway willows, the glistering currents bounding along like a dozen otters seizing ownership of the place, as we walk within sight. Our bodies, 70 percent water (and our brains more), only mimic the earth's surface in this respect. And we want a mixed and muscular sky, bulging yet depthless, and full of totems, talismans, in the clouds: not every day but when we have the energy for it, just to know that we're alive. Rising land of course will lift our spirits too. Hills, a ridgeline, not to begin toiling right up today but the possibility of doing so, perhaps discovering unmapped crannies up there and trees as tiny as bonsai on the crest, yet dips for the eyes to rest in as we look.

We already think we know too much about too much, so mountains are for the mystery of ungeometric convolutions, a boost without knowing what's on top. Awe is not a word much used lately, sounding primitive, like kerosene lamps. What's to be awed about— is this the Three Wise Men following the Star?—what hasn't been explained? Actually, I don't know what has been explained. If we are told, for example, that 99 percent of our genes are similar to those of a mouse, does that explain anything? Apprehension, disillusion, disorientation, selfishness, lust, irony, envy, greed, and even self-sacrifice are commonplace: but awe? Society is not annealed enough. Trust and continuity and leadership are deteriorating, and the problem when you are alone is the clutter. Finding even a sight line outdoors without buildings, pavement, people, is a task, and we're not awed by other people anymore: too much of a good thing. We need to glimpse a portion of the axle, the undercarriage, of what it's all about. And mountains (an axis, if not an axle) are harder to be glib about than technological news reports. But if you wait until your mature years to get to know a patch of countryside intimately, thoroughly, your responses may be generic, not specific—just curiosity and good intentions—and you will wind up going in for golf and tennis and power mowers, bypassing nature instead.

NO MAN WAS COMPLETE without a parrot on his shoulder, I used to think. Pirates had them—or perhaps a monkey with a string knotted around its waist—and far-flung sailors, and naturalists searching the tropics for undiscovered plant and animal species. An Orinoco toucan or an orange-epauletted Amazon or hyacinth macaw nibbled at an earlobe or chatted in their ear. At the mouth of the Congo River or the Amazon, hotels had to post a sign saying no parrots were allowed here, and the birds lived so long that in tamer harbors like the USA you might never know who had taught yours to cry, What's that down between your legs, big boy? In the port area of Lower Manhattan which later became the World Trade Center, I used to see foulmouthed merchant seamen's big-billed birds for sale in a cigar store that also proffered shrunken human heads with pained and puckered faces and sewn-up lips which sailors on the coffee or United Fruit banana boats had purchased from tribes such as the famous head-hunting Jivaros of Ecuador. Both the brilliant-colored parrots and the Indians' heads, suspended behind the counter by their greasy black braids of hair, had been jungle-born, except of course for the especially valuable blonde-tressed heads of white women and their missionary husbands: although, buyer beware, you were supposed to be careful about fakes—maybe monkeys that had been treated and bleached.

And there was a kind of "leopard store," as I thought of it, named Trefflich's, in a brownstone at 215 Fulton Street, close to where a lot of other ships came in to dock from Joseph Conrad countries. It sold jaguar cubs, anacondas, margay cats and ocelots, aoudads and addaxes, baboons, pangolins, gibbons, adolescent elephants—importing wholesale stuff for zoos to a warehouse in New Jersey. But you could walk around the several floors, if you were with your father, and look at giant Seychelles tortoises, reticulated or Burmese pythons, black panthers peering between the slats of cargo crates, and wheedling monkeys whose organ grinder might have died. Carnival owners stopped by in the spring in painted trucks to purchase an iron cage with a sun bear already in it or rent a half-trained lion, or a bunch of monkeys. "When it comes to monkeys," a placard boasted, "we pledge ourselves to give full cooperation to all operators interested in giving the public their monkey's worth!" Beasts in makeshift confinement—an arctic wolf, a rainbow boa, a baby camel— crammed every corner, and then in season might be touring the nation's midways, living on roadkills or sick chickens the drivers stopped to pick up, the panther on foundered horses or dead dogs from a pound, the monks on fruit the public bought.

Parrots did not remain a priority for me because I sensed that they were delicate and in considerable peril, though squawking harshly and nipping fingers. Even when a fancier hamstrung them by scissoring their flight feathers, as if to baubleize them, they continued to emit untamable screams and like a peg-leg pirate moved about laboriously by grabbing footholds with their beaks and chimneying up or belaying down in mountaineer fashion. Their shrieks might bring the neighborhood's blue jays to the owner's window as if to try to help a friend in need, and double the noise. Then you'd see the guy abandon a thousand-dollar pet to his local flower shop, where at least it had the ferns and ficus trees for company. Or I've known a parrot or two that escaped from captivity and shimmied high into a fir tree next to the house, and even in the wintertime simply refused to come down. Up, up, the pinioned bird hitched with claws and beak, watching the hollering jays and crows circling around and screaming gleefully with them. Although of course it couldn't fly, it ate a few tart bits of bark or cone in freedom at their level. The drama continued for hours— pleas and commands from the ground, and hullabaloo from the whirling wild birds. Then a soft snow started falling, as night settled down. The native flocks—warmly plumaged and observing the newcomer's crippled condition—flew away to their sheltered roosts, while the parrot, in its bright jungle colors, climbed poignantly, stoically higher, to wait in silence to nibble needles and freeze.

I went to summer camp in the Adirondacks, helped out at an animal hospital near my home, and, with a friend's family, visited a little dude ranch in the Wind River range at Dubois, Wyoming, going out by train when I was fourteen. This showed me that whole tiers of land exist that most of us never reach; just look at, perhaps. My horse could scramble by switchbacks for a short way, like a badger galumphing. Horses were more independent-minded than a dog, preferring the open range as a grubstake to any barn and wintering there for six months as uncosseted as the elk or mule deer. The ranch hands wintered pretty tough, too—not just drank a lot but practically hibernated in snowed-over cabins, living on a wad of cash secreted in a coffee can, not a bank account, and snaring jackrabbits, eating root-cellar turnips and steaks axed off a frozen side of beef, by a hissing Coleman lantern. They lost their teeth sooner than Easterners, and the men got gimpy at an early age from being thrown or kicked. Not only the rodeo types: many ordinary wranglers were fallen on, stepped on, in breaking horses, roping calves, rassling a steer, or had slid off an icy road in a Chevy and limped for miles with a broken bone. Help was so damned far off.

The bristling, pelagic scale of the landscapes, skyscapes, exhilarated me, plus the chance to catch sight of a cougar by peering up a box canyon, or the coyotes that howled after dark from the same creek bed where I had walked an hour before. In these late 1940s the Good War was barely over, evil had been defeated, but a tremor of risk and early death still prickled the mood of many people of middle age: veterans who wouldn't speak of what they'd done, and for whom foreign travel had involved stifling weeks on a troop ship to places they never wanted to lay eyes on again. They squinted and bit their lips, thinking back. Fred, the one cowhand I got to know somewhat—who taught me to ride in a roundup and shoot ground squirrels, and talked confidingly toward the end—was both mild in manner and steadfast, yet lamed internally, with a cowboy's kidney disabilities and a flat-wallet winter to look ahead to. As with Hope, I can't reconstruct our conversations from sixty years ago, except that there were many factualities for me to learn, glued to yearnings Fred might even help me fulfill. Our talk was less inchoate. Childless from knocking around the

West all his life, he was sympathetic to my wish to hear stories of mega-wildlife, trapper-hermits, gold prospecting, bigfoot myths, and not just rehearse my saddle skills like the other dudes. That West was already threadbare but not skeletal, and I learned that when somebody in the know recognizes what you care about, he may earnestly try to help. Antelope, moose, and marmots—"whistlers"—we looked for, and falling-down cabins in draws that had yams attached to them. Fred was slightly built, like a person who dealt with creatures so large that heft itself hardly mattered, compared with logic and telepathy. He sized up people quietly too, and minimized his reactions if he could, the way you would with a haywire heifer or bull.

Act purposefully but minimally and keep your reasons under wraps, was a lesson he taught me. Not the whole formula for life, but quite a beginning, because love and openness to what you love are fragile and yet will flower if cupped and sunlit: as will a freelance toughness and survivability, when you need that. Like a certain helicopter pilot in the Brooks Range in Alaska who flew me around decades later, Fred grounded my enthusiasms at the same time that perhaps mine reinforced his. I couldn't help him face an old age of penury, but we were wistful when this summer interlude wound up. Teasing so many memories out of his mind had cheered him up, made him feel that they were worth it, and as in a relay race, he was passing along nuggets to me, not necessarily from his own life but that turn-of-the-century horse wranglers had conveyed to him. Just so, we elasticize our lives—as you'll see a tiny school of fish do in an aquarium. As quick as mercury and multidirectional, they impart a darkly silver, wriggly sorcery to the cubic inches of the tank. Instead of gallons, it becomes like having a miniplanetarium in the house, because the stars also sometimes seem to swim in the sky, not just hang in suspension there.

Pets in containers, or loose as catty companions or doggy slaves, can hardly fill in for the immensity of wind, stars, and trees, the infinity of unlobotomized animal species, the intricacy of landscapes, the galaxy of scents and shapes in natural creation, that we are losing, or just no longer sense or see. A planetarium is not the heavens, or an aquarium the southern oceans, and our own intricacy—our bristly whiskers, flaring nostrils, our fingerpads flicking in and out as ceaselessly as gills, our curling pinnae and peripheral vision and intuitive antennae, all seeking connections—perhaps demands them. Only 2 percent of Americans are farmers now, and yet the rest of us are still avid for spring's green-up and weather forecasts. Without the primeval dangers that formed us, we tune in bruising professional football or pore through the tabloids for raunchy murders, sexual triangles and kidnappings, news of disease, greedy scandals. We actually learn skills of the chase and the feint from these, learn about insanity and bad judgment and to control our spates of rage, to cushion our marriages, downsize our fantasies, put the brakes on our Neanderthal instincts. The tabloids are appetite-rich and Darwinian. We read them for meat and war games, or watch the tube for boobs that we can't ogle in real life, and truth or consequences—robbers punished—while rejiggering our minds' chemistry with pills, replacing an aging hip with titanium, and exercising on a gym machine, or face-lifting our long-suffering skin.

But I seem to have gained, around eight, nine, or ten, the rather precious sense of continuity that knows that when you come out of the woods into a house it's only temporary; you will be going back out again. People are less amphibious or ambidextrous in this regard than they used to be. A thousand or so may have topped Mt. Everest ("Well we knocked the bastard off," Sir Edmund Hillary famously said, after conquering it in 1953), and plenty run marathons or balance on surfboards. Yet a more authentic affinity with what we call nature is being lost even faster than nature itself. Into the void slips obsessional pornography, fundamentalist religion, strobe-light showbiz (no Bing Crosby or Frank Sinatra, who blazed on forever), and squirmy corporate flacks such as the old power brokers seldom employed. If gyms don't substitute for walking, it's hard to find a place to walk, as houses line every beachfront and scissor every patch of woods with cul-de-sacs for real estate. You may prefer the ubiquity of electricity to seeing fields of stars after dark, but losing constellation after constellation in the night, and countless water meadows along uncontoured rivers, and bushy-tailed horizons, may be a titanic change. Our motors similarly wipe out the buzz and songs of insects, birds, the sibilation of the breezes that hunters used to front, always stalking into the wind and studying the folds of the terrain for how it flowed, because meals were won by knowing the intimacies of the wind. To lose moonlight, and compass placement, and grasshoppers telling us the temperature by the intensity of their sound, poses the question of whether we can safely do away with everything else. The ecology of solitary confinement on this planet may be calamitous: not to mention the sadness. To assuage the emotional effects, already one notices an explosion of plant nurseries, pet stores, computer-simulated androids, and television animations. We've boarded up our windows so as to live interiorly with just our own inventions—though sensing too that we are in the grip of a slow, systemic illness, somehow pervasive—as meanwhile chimpanzees are being eaten up wholesale in Africa as "bushmeat," the elephants butchered, the lions poisoned.

I knew these signature animals by the age of eighteen because I worked for two summers in a circus where we had in our charge some of the most glorious and legendary wilderness creatures. Asian elephants, Sumatran and Siberian and Bengal tigers, a cheetah, a hippo, a jaguar, pythons and boas, three lowland gorillas, a rhino, and an orangutan. My mentor then was a Mohawk Indian, from a culture that was comparably endangered. Indeed, he finished out the remainder of his years as a groom at the riding stable that services New York's yuppie equestrians who ride in Central Park, before having his ashes scattered off of the George Washington Bridge into the Hudson River, which are immemorial waters for the Mohawks. But the survival of wild places and wild things, like the permanence of noteworthy architecture, or the opera, a multiplicity of languages, or old shade trees in old neighborhoods, is not a priority for most people. They are on their way out, and you simply love and love them as you, too, shuffle along.

But the elephants, wrinkled in their sagging hides, appeared to recognize the tenor of events. They were not optimistic—at least I thought not—and forty years later, seeing shattered herds in India and Africa, I was surer still that they realized that the road for them shambled off downhill. Their anxiety was more than jumpy; it seemed demoralized. Their bizarre hugeness only doomed them further. We generally discover important things late: like how very closely the great apes' genome resembles ours. This was obvious to the naked eye and won't prevent wild ones from being eaten in Africa, but makes their treatment here in captivity more appalling. And thus it was with elephants' infrasonic communications, which supplement the squeals and trumpeting we hear. It took a former whale biologist, Katie Payne, who had helped record the high frequencies humpbacks sometimes use, beyond the capacity of a human ear, to figure out that elephants also talk at acoustic (though subsonic) levels we can't detect, and that these deep sound waves travel as messages for surprising distances, from herd to herd (yet nothing like what the ocean's physics can accomplish for certain whales' low-voiced emotings).

THE MORE COMPLEXITIES we come to know about a fellow being, the less cavalier we're going to feel when its kin are wiped out. Most species that disappear, of course, have never been examined or "discovered" at all. But with the jumbo kind—formerly demonized as rogues, or boat swampers and living oil wells—we have a good deal less excuse. Indeed, in the circus, decades before Katie Payne's breakthrough, I had experienced intimations that within our single herd, animals a hundred yards apart could convey their politics or frustrations by sounds below the lowest range that people heard. They would be clearly communing across the field, looking at each other, swinging their trunks convivially and swaying with eloquent body language, until after a minute or so the session ended with a strain of sound finally edging up into a low-pitched groan. I was eighteen, nineteen, not a scientist, and these insights were accompanied by a swarm of others about our giant, protean, poignant beasts—Ruth and Modoc and twenty others—whose feet I liked to lie close to, testing my trust in the rapport I thought I had with their rhythms and whims. Acoustics were not the reason I was touched or central to what I was trying to comprehend, even when they stood there forthrightly and frontally, broadcasting sounds I sometimes intuited but couldn't hear.

In East Africa on two trips during the 1970s, I saw pristine herds on the vastness of the veldt, browsing slowly among the thorn trees as creatures do when engaged in being themselves. Although they were being poached, the horizons were huge, and the scale of ivory-hunting an attrition they could bear. Their humor, gait, and dignity were intact, the tutoring of the calves, the playfulness of bathing, the virtuoso trunks spraying dust when insects annoyed them, or plucking an epicurean shoot, or squealing at a stork. By the 1990s, however, when I returned twice again, the splintered groups, targeted by Somalis with Kalashnikovs, had witnessed so much butchery and anguish—their numbers more than halved—that they acted as if danger were everywhere. They drank at the water holes twitchily, hastily, and migrated between their feeding groves without the ambling ambience of old. Like the chimps I saw, they didn't just react to immediacies, as, for instance, the big cats did, but appeared to worry in advance. They weren't freewheeling personalities anymore, and it was a relief to meet a noncommittal aardvark or a snoopy jackal on the track.

Nature throbs in us through our digestive gases, sweaty odors, wrist pulse, unruly penis or bloody vulva, and nervy tics. We flinch, gasp, fuck, cluck, grin, blink, panic, run, fight, sleep, wake, and wolf a meal like animals. Our official seven deadly sins are rather animal, too, and so is bliss, I think: not only lust but that out-of-body happiness you may feel when being quite still, yet aware and self-contained. Nature is continuity with a matrix and not about causing a stir in the world, and as we destroy our links to other forms of life, it's like whittling at our heels and shins and toes. You can do it for a while until you cut a tendon, nick a bone, and find you limp. And we've now done that. Life turns into more of a riddle when not braided together with other manifestations of energy, grace, scale, and harmonics or tempo and all the rest. Humanity all alone can be constricting, and I've met more blithe spirits in frontier situations than anywhere else. They weren't the quickest conversationalists or most educated, and inevitably there were also augured souls who had fled to the mountains to get as far away from other people as they could. But as George Orwell remarked at the end of his diaries, "At fifty, everyone has the face he deserves" (alas, he didn't live that long); and these guys from the era of World War I or the Depression, living on the Skeena River or the Stikine, in British Columbia and Alaska—the Spatsizi or the Omineca, the Klappan or the Kuskokwim, the Tanana or the Porcupine—when you hollered to them from the footpath and they came out of their cabins, looked blithe. They were likely to wear long johns all year round, and light the woodstove every morning regardless of how warmly the sun might be going to shine, because you never know. Anyway, smoke and long johns discourage mosquitoes, and if you've ever been profoundly cold you won't mind being over-warm.

They were on the lookout for gold colorations in the creek beds as we walked about, and before fall got well started they would be laying in a mammoth woodpile, and extra rations under the floorboards, and boiling and re-scenting their fur traps, then, after a hard frost, throwing caribou carcasses up on the pitched roof, where they'd keep. I made sure not to take them unawares—because was this an individual who lived out here with the elements because he had abandoned everybody who had ever trusted him, or because of what he'd sought? With men thirty years older than me and at home with the tessitura of the wilderness in the 1960s, you didn't need to be a psychologist to arrive at some swift conclusions, mostly cheerful ones. A general competence, plus maybe the yearly salmon runs, had enabled them to ride out the six months of winter, as well as the specters that can afflict a solitary mind. If a man's smile looked to be guyed out as securely as a well-staked tent, it meant he probably grinned a lot, if only to himself, and wasn't about to blow away.

My hunches in the main worked out, and more important I escaped the confines of my stutter and gradually became able to talk to people as easily as to animals. Being a humanist, I was not as interested in animals but, in the Whitman mode, aspired to contain multitudes, which included being a mutt and hybrid oneself, snaky, fishy, foxy, and as Afro-Indo-European as our far-trekking forefathers. Although I was living in New York City by this time and married, and therefore swinging as if in a bathysphere into some of these roadless valleys in the far Northwest with my pack and sleeping bag, I felt at home on a moose path too. My spirit keyed into the tuning fork of old melodies—not simply the sense of trust I had acquired in Dr. Green's woods, but the narratives, I think, behind us all. In back of Gilgamesh and Beowulf, Homer, Hardy, and Melville, lie impulses of animism that personified the sweet wellsprings and ominous cliffs, the mountaintops and antique trees as godlings for our ancestors, and the fact that primates are talking to lions even today. Baboons are arguing with them on a Tanzanian plain and fathoming their reactions much as I was often doing half a century ago in the Ringling Bros. Circus. We sleep in edgy surges of a few hours that we manage to combine into a night's civilized schedule, yet would have more logic for a chimp in the forest or a baboon on the veldt, rousing from each nap to glance around for a leopard's dappled coat creeping through the gloaming.

Life is so elastic that people whose circumstances appear to be about the same may measure themselves as almost anywhere on a continuum from misery to elation, and nature herself is invoked to justify fidelity or infidelity, tolerance or violence. I've never thought of nature as a guardian angel, but rather as a polychromatic thrum you sway and hum along with and therefore are not caught by surprise by a sudden juddering in the weather, a hyena on a kill, or a soggy bowl of landscape, when you're hiking. Although never a daredevil, I didn't believe that we can live quite wholeheartedly if we are overly afraid of dying, but, on the other hand, didn't think if life gets boring you have to climb a hairy mountain. Just pick your calluses off and refresh your sensibilities. The airiest scenery I've been privileged to see was in the Himalayan foothills of Arunachal Pradesh, between Assam and Tibet in northeast India. Yet I didn't stand there yearning to scale and "top" the greenly rising and then vertiginous ridges that towered toward snow peaks like laddered but amorphous ghosts, muscular and portentous beyond the mists. I wanted to let them be. And similarly I plan not to be cremated, so that the proverbial worms can do a recycling job on me after I die, rather than be rendered into tidy, sterile minerals in an undertaker's furnace: a less juicy fate.

Now, animals live even more in the present than we do. They are geographic or hierarchical in organization, operating by rote or scented memories of previous hazards and good fortune, seeking food with smaller brains but not wistful about it, as you'll notice watching a fox glance up at suet in a birdfeeder without wasting energy in pining after a bite of it. Short and brutish has been a description of their lives ("brutish" being somewhat tautological), but certainly the lives of what are called the megafauna are getting shorter while ours grow longer. Some people scarcely know what to do with their bonus time—doubled life spans, plus the round-the-clock availability of artificial light—because nature doesn't deal in bonuses. The sun rises and sets when it did a million years ago, with daylight altering by immemorial increments as the planet rolls. It doesn't award you an extra hour if you have a deadline. Can you make it? nature asks instead, if it says anything at all. But secondly, and curiously, I think, it speaks in terms of glee. Glee is like the froth on beer or cocoa. Not especially necessary or Darwinian, it's not the carrot that balances the stick, because quieter forms of contentment exist to reward efficiency. Glee is effervescence. It's bubbles in the water—beyond efficiency—which your thirst doesn't actually need.

Bubbles are physics, not biology, and glee, if the analogy is to carry far, may be an artesian force more primordial than evolutionary. To me, it's not a marker for genetic advantages such as earning more, but an indicator that life—the thread of Creation, the relic current that has lasted all this way—is ebullient. Still, you might argue that the choosiest females select not just for strength and money or its zoological equivalents, but for the superfluous energy that humor and panache imply. The woodpecker drumming an irregular tattoo on my tin roof in the spring is not mechanistic in his ritual, as if merely to prove that he could dig big bugs out of a tree and bring them to his mate. His zest and syncopation is like when you watch two fawns gamboling with a doe, or a swaggering vixen mouth three meadowmice that she has killed to fit them all between her teeth for the trip to her den. Such surplus moments relax us and serve a tonic function—triumphal for the vixen, toning the fawns' reflexes, letting the woodpecker pause unexpectedly to listen for an answer.

The gamboling, like a kitten's stalk, prepares an animal for the hunt or being hunted, and the youngster that enjoys it most may wind up savviest. But the glee I mean is less utilitarian, more spontaneous, and a kind of elixir that needs a bit of peace to germinate. How does one account for the passive, concentrated happiness of listening by a lake to the lap and hiss of rustling water, watching the leaves jiggle, the poplars seethe and simmer. The lake is ribbed with ropes of wind and strands of sun between cloud shadows. The contours of three hills delineate the comely way that brooks feed into its blue bulk, and otters, loons, mergansers, animate it (the far mile curving out of sight), so that you'd hardly need to invent a loch "monster" for drama. And yet you can wake up nearly anywhere and experience a comparably high-pitched serenity. Glee is not complacency—in the middle of a roaring city it may seize you—and I think of it as possibly generated at life's origins, like a filament from, or footprint of, that original kick. Nature seems more than Evolution, punctuated or otherwise, and the Creationists may be onto something when they insist that it is an effusion of God's glory. Their god isn't mine, but glee may be a shard of divinity.

Nature, although more inclusive than fundamentalism allows for, seems to me infused with joy. Even the glistering snow is evidence, though burdensome by March, and October's dying leaves, parched by an internal trigger before the first frosts, turn gratuitously orange, red, and yellow, as beautiful as any plumage—yet what mating purpose does that serve? When outdoors with a dog, anybody can observe the gulping relish with which it quaffs evocative smells, then punctiliously may leave its own before hounding on. I have been watching colonies of animals, from chickens, mice, and garter snakes to some of the megafauna, for sixty years, and when they are not under stress you see plenty of delight and exuberance, particularly when young ones are splitting off and diligently getting a new group started. Biochemistry drives hunger and explains why animals consume one another. But what explains the elation, exuberance—this surplus snap of well-being that animals as well as naturalists feel, and people in Calcutta as much as in New York, or Arunachal Pradesh, for that matter?

Joy sprouts from squalor as well as in the middling classes, a perennial as well as a primitive emotion, as if propelled by a spin originating from the ur- or ultra-density of the Big Bang. Or should we claim that amphibians only acquired a capacity for glee after they became lizards? Or lizards only after they evolved into birds? Where and when did the perception of beauty begin? Most of us nowadays agree that the birds that sing at dawn in the spring are expressing some degree of gladness in their surging notes, not merely a mechanical territoriality. But for a person like me who considers the toads' sparkling, twinned-note, extended song on warm days in May and June to be actually the loveliest of all, the answer is not that easy. I can't swallow the notion that I—but not the toads—find it so lovely. (I also think I've seen and heard alligators and seen turtles enjoy themselves.) However, then the question shifts to whether amphibians that sing, such as frogs and toads, only began to respond to warmth and what we call beauty after they left the constancy of the water and ceased being fish. Not a sure-shot answer there either, unless you discount the evidence of your eyes when you're closely watching fish. And water is an unboxed, undulant medium. What does it mimic when it sloshes?

That crucial age when I opened up and trusted myself to nature, back in Connecticut in 1942, is about when most children start perceiving the world beyond themselves in nuanced, revelatory ways. I later tried sport hunting and gave it up, sport climbing and gave it up, preferring not to lord my ego over what I saw, as in those chill, steep rain forests of Arunachal Pradesh—like Alaska's, multiplied several times over—with footbridges woven of vines stretching across the cataracts and thatched houses perched on stilts and white peaks suspended above it all. To be immersed was sufficient, without attempting to "knock the bastard off."

If we're not immersed, we're likely to try to simulate the hubbub of a tribal encampment by collecting cats or dogs (butchering countless horses in order to feed them), or barbecue sets and fishing tackle, off-road vehicles or quirky website monikers. We'll fly in bales of greenhouse flowers from low-wage tropical countries, which are being denuded of their natural flora, to present as symbols of we hardly know what. That is, yes, for anniversaries, marriages, courtship, holidays, graduations: but why flowers? Are we bees or bears, or are they somehow akin to the mysteries of glee and orgasm and why small boys stand by the conundrums embedded in the mud of a pond, then reach from the bank or roll their pant legs up and wade after salamanders, water snakes, pollywogs, and perhaps a reflection of what have you?

We reach for where we came from, our older folk a bit homesick: the nights not being starry anymore and distances not quite real. Is there anything untoward that we don't take a pill or press a button for? Nature envelops us, nonetheless, in the piquancy of cottage cheese, the giggle of thunder in the next county. Our lewdness and acquisitiveness bray to prove how recidivist we are, still with our feet in the primal muck. I live alone at the moment, and would smell piquant after a stroke, if I weren't discovered immediately. Nor, when I laugh, do I feel in the twenty-first century—I could be Babylonian. And my rapport with friends is more a refinement of ancient habituations than contemporary. Nature, when abused, may react eventually like a tiger whose tail has been pulled. We shall see, indeed, if that is the case. We will definitively find out. But in the meantime we live like those amphibians: sometimes on the dry beach of modernity and sometimes swimming in the oceans that were here eternally before.

DAVID QUAMMEN

Darwin or Not

FROM *NATIONAL GEOGRAPHIC*

It may seem startling that a major magazine today would need to defend the validity of evolutionary theory. But given that nearly half of Americans disavow evolution, it's no wonder that the editors of National Geographic *commissioned nature writer David Quammen to utter a hearty "No" to the question "Was Darwin wrong?"*

Evolution by natural selection, the central concept of the life's work of Charles Darwin, is a theory. It's a theory about the origin of adaptation, complexity, and diversity among Earth's living creatures. If you are skeptical by nature, unfamiliar with the terminology of science, and unaware of the overwhelming evidence, you might even be tempted to say that it's "just" a theory. In the same sense, relativity as described by Albert Einstein is "just" a theory. The notion that Earth orbits around the Sun rather than vice versa, offered by Copernicus in 1543, is a theory. Continental drift is a theory. The existence, structure, and dynamics of atoms? Atomic theory. Even electricity is a theoretical construct, involving electrons, which are tiny units of charged mass that no one has ever seen. Each of these theories is an explanation that has been confirmed to such a degree, by observation and experiment, that knowledgeable experts accept it as fact. That's what scientists mean when they talk about a theory: not a dreamy and unreliable speculation, but an explanatory statement that fits the evidence. They embrace such an explanation confidently but provisionally—taking it as their best available view of reality, at least until some severely conflicting data or some better explanation might come along.

The rest of us generally agree. We plug our televisions into little wall sockets, measure a year by the length of Earth's orbit, and in many other ways live our lives based on the trusted reality of those theories.

Evolutionary theory, though, is a bit different. It's such a dangerously wonderful and far-reaching view of life that some people find it unacceptable, despite the vast body of supporting evidence. As applied to our own species, *Homo sapiens*, it can seem more threatening still. Many fundamentalist Christians and ultra-orthodox Jews take alarm at the thought that human descent from earlier primates contradicts a strict reading of the Book of Genesis. Their discomfort is paralleled by Islamic creationists such as Harun Yahya, author of a recent volume titled *The Evolution Deceit*, who points to the six-day creation story in the Koran as literal truth and calls the theory of evolution "nothing but a deception imposed on us by the dominators of the world system." The late Srila Prabhupada, of the Hare Krishna movement, explained that God created "the 8,400,000 species of life from the very beginning," in order to establish multiple tiers of reincarnation for rising souls. Although souls ascend, the species themselves don't change, he insisted, dismissing "Darwin's nonsensical theory."

Other people too, not just scriptural literalists, remain unpersuaded about evolution. According to a Gallup poll drawn from more than a thousand telephone interviews conducted in February 2001, no less than 45 percent of responding U.S. adults agreed that "God created human beings pretty much in their present form at one time within the last ten thousand years or so." Evolution, by their lights, played no role in shaping us.

Only 37 percent of the polled Americans were satisfied with allowing room for both God and Darwin—that is, divine initiative to get things started, evolution as the creative means. (This view, according to more than one papal pronouncement, is compatible with Roman Catholic dogma.) Still fewer Americans, only 12 percent, believed that humans evolved from other life-forms without any involvement of a god.

The most startling thing about these poll numbers is not that so many Americans reject evolution, but that the statistical breakdown hasn't changed much in two decades. Gallup interviewers posed exactly the same choices in 1982, 1993, 1997, and 1999. The creationist conviction—that God alone, and not evolution, produced humans—has never drawn less than 44 percent. In other words, nearly half the American populace prefers to believe that Charles Darwin was wrong where it mattered most.

Why are there so many antievolutionists? Scriptural literalism can only be part of the answer. The American public certainly includes a large segment of scriptural literalists—but not *that* large, not 44 percent. Creationist proselytizers and political activists, working hard to interfere with the teaching of evolutionary biology in public schools, are another part. Honest confusion and ignorance, among millions of adult Americans, must be still another. Many people have never taken a biology course that dealt with evolution nor read a book in which the theory was lucidly explained. Sure, we've all heard of Charles Darwin, and of a vague, somber notion about struggle and survival that sometimes goes by the catchall label "Darwinism." But the main sources of information from which most Americans have drawn their awareness of this subject, it seems, are haphazard ones at best: cultural osmosis, newspaper and magazine references, half-baked nature documentaries on the tube, and hearsay.

Evolution is both a beautiful concept and an important one, more crucial nowadays to human welfare, to medical science, and to our understanding of the world than ever before. It's also deeply persuasive—a theory you can take to the bank. The essential points are slightly more complicated than most people assume, but not so complicated that they can't be comprehended by any attentive person. Furthermore, the supporting evidence is abundant, various, ever increasing, solidly interconnected, and easily available in museums, popular books, textbooks, and a mountainous accumulation of peer-reviewed scientific studies. No one needs to, and no one should, accept evolution merely as a matter of faith.

TWO BIG IDEAS, NOT JUST ONE, are at issue: the evolution of all species, as a historical phenomenon, and natural selection, as the main mechanism causing that phenomenon. The first is a question of what happened. The second is a question of how. The idea that all species are descended from common ancestors had been suggested by other thinkers, including Jean-Baptiste Lamarck, long before Darwin published *The Origin of Species* in 1859. What made Darwin's book so remarkable when it appeared, and so influential in the long run, was that it offered a rational explanation of how evolution must occur. The same insight came independently to Alfred Russel Wallace, a young naturalist doing fieldwork in the Malay Archipelago during the late 1850s. In historical annals, if not in the popular awareness, Wallace and Darwin share the kudos for having discovered natural selection.

The gist of the concept is that small, random, heritable differences among individuals result in different chances of survival and reproduction—success for some, death without offspring for others—and that this natural culling leads to significant changes in shape, size, strength, armament, color, biochemistry, and behavior among the descendants. Excess population growth drives the competitive struggle. Because less successful competitors produce fewer surviving offspring, the useless or negative variations tend to disappear, whereas the useful variations tend to be perpetuated and gradually magnified throughout a population.

So much for one part of the evolutionary process, known as ana-genesis, during which a single species is transformed. But there's also a second part, known as speciation. Genetic changes sometimes accumulate within an isolated segment of a species, but not throughout the whole, as that isolated population adapts to its local conditions. Gradually it goes its own way, seizing a new ecological niche. At a certain point it becomes irreversibly distinct—that is, so different that its members can't interbreed with the rest. Two species now exist where formerly there was one. Darwin called that splitting-and-specializing phenomenon the "principle of divergence." It was an important part of his theory, explaining the overall diversity of life as well as the adaptation of individual species.

This thrilling and radical assemblage of concepts came from an unlikely source. Charles Darwin was shy and meticulous, a wealthy landowner with close friends among the Anglican clergy. He had a gentle, unassuming manner, a strong need for privacy, and an extraordinary commitment to intellectual honesty. As an undergraduate at Cambridge, he had studied halfheartedly toward becoming a clergyman himself, before he discovered his real vocation as a scientist. Later, having established a good but conventional reputation in natural history, he spent twenty-two years secretly gathering evidence and pondering arguments—both for and against his theory—because he didn't want to flame out in a burst of unpersuasive notoriety. He may have delayed, too, because of his anxiety about announcing a theory that seemed to challenge conventional religious beliefs—in particular, the Christian beliefs of his wife, Emma. Darwin himself quietly renounced Christianity during his middle age, and later described himself as an agnostic. He continued to believe in a distant, impersonal deity of some sort, a greater entity that had set the universe and its laws into motion, but not in a personal God who had chosen humanity as a specially favored species. Darwin avoided flaunting his lack of religious faith, at least partly in deference to Emma. And she prayed for his soul.

In 1859 he finally delivered his revolutionary book. Although it was hefty and substantive at 490 pages, he considered *The Origin of*

Species just a quick-and-dirty "abstract" of the huge volume he had been working on until interrupted by an alarming event. (In fact, he'd wanted to title it *An Abstract of an Essay on the Origin of Species and Varieties Through Natural Selection*, but his publisher found that insufficiently catchy.) The alarming event was his receiving a letter and an enclosed manuscript from Alfred Wallace, whom he knew only as a distant pen pal. Wallace's manuscript sketched out the same great idea—evolution by natural selection—that Darwin considered his own. Wallace had scribbled this paper and (unaware of Darwin's own evolutionary thinking, which so far had been kept private) mailed it to him from the Malay Archipelago, along with a request for reaction and help. Darwin was horrified. After two decades of painstaking effort, now he'd be scooped. Or maybe not quite. He forwarded Wallace's paper toward publication, though managing also to assert his own prior claim by releasing two excerpts from his unpublished work. Then he dashed off *The Origin*, his "abstract" on the subject. Unlike Wallace, who was younger and less meticulous, Darwin recognized the importance of providing an edifice of supporting evidence and logic.

The evidence, as he presented it, mostly fell within four categories: biogeography, paleontology, embryology, and morphology. Biogeography is the study of the geographical distribution of living creatures— that is, which species inhabit which parts of the planet and why. Paleontology investigates extinct life-forms, as revealed in the fossil record. Embryology examines the revealing stages of development (echoing earlier stages of evolutionary history) that embryos pass through before birth or hatching; at a stretch, embryology also concerns the immature forms of animals that metamorphose, such as the larvae of insects. Morphology is the science of anatomical shape and design. Darwin devoted sizable sections of *The Origin of Species* to these categories.

Biogeography, for instance, offered a great pageant of peculiar facts and patterns. Anyone who considers the biogeographical data, Darwin wrote, must be struck by the mysterious clustering pattern among what he called "closely allied" species—that is, similar creatures sharing roughly the same body plan. Such closely allied species tend to be found on the same continent (several species of zebras in Africa) or within the same group of oceanic islands (dozens of species of honeycreepers in Hawaii, thirteen species of Galápagos finch), despite their species-by-species preferences for different habitats, food sources, or conditions of climate. Adjacent areas of South America, Darwin noted, are occupied by two similar species of large, flightless birds (the rheas, *Rhea americana* and *Pterocnemia pennata),* not by ostriches as in Africa or emus as in Australia. South America also has agoutis and viscachas (small rodents) in terrestrial habitats, plus coypus and capybaras in the wetlands, not—as Darwin wrote—hares and rabbits in terrestrial habitats or beavers and muskrats in the wetlands. During his own youthful visit to the Galápagos, aboard the survey ship *Beagle,* Darwin himself had discovered three very similar forms of mockingbird, each on a different island.

Why should "closely allied" species inhabit neighboring patches of habitat? And why should similar habitat on different continents be occupied by species that aren't so closely allied? "We see in these facts some deep organic bond, prevailing throughout space and time," Darwin wrote. "This bond, on my theory, is simply inheritance." Similar species occur nearby in space because they have descended from common ancestors.

Paleontology reveals a similar clustering pattern in the dimension of time. The vertical column of geologic strata, laid down by sedimentary processes over the eons, lightly peppered with fossils, represents a tangible record showing which species lived when. Less ancient layers of rock lie atop more ancient ones (except where geologic forces have tipped or shuffled them), and likewise with the animal and plant fossils that the strata contain. What Darwin noticed about this record is that closely allied species tend to be found adjacent to one another in successive strata. One species endures for millions of years and then makes its last appearance in, say, the middle Eocene epoch; just above, a similar but not identical species replaces it. In North America, for example, a vaguely horselike creature known as *Hyracotherium* was succeeded by *Orohippus,* then *Epihippus,* then *Mesohippus,* which in turn were succeeded by a variety of horsey American critters. Some of them even galloped across the Bering land bridge into Asia, then onward to Europe and Africa. By five million years ago they had nearly all disappeared, leaving behind *Dinohippus,* which was succeeded by *Equus,* the modern genus of horse. Not all these fossil links had been unearthed in Darwin's day, but he captured the essence of the matter anyway. Again, were such sequences just coincidental? No, Darwin argued. Closely allied species succeed one another in time, as well as living nearby in space, because they're related through evolutionary descent.

Embryology too involved patterns that couldn't be explained by coincidence. Why does the embryo of a mammal pass through stages resembling stages of the embryo of a reptile? Why is one of the larval forms of a barnacle, before metamorphosis, so similar to the larval form of a shrimp? Why do the larvae of moths, flies, and beetles resemble one another more than any of them resemble their respective adults? Because, Darwin wrote, "the embryo is the animal in its less modified state" and that state "reveals the structure of its progenitor."

MORPHOLOGY, HIS FOURTH CATEGORY OF EVIDENCE, was the "very soul" of natural history, according to Darwin. Even today it's on display in the layout and organization of any zoo. Here are the monkeys, there are the big cats, and in that building are the alligators and crocodiles. Birds in the aviary, fish in the aquarium. Living creatures can be easily sorted into a hierarchy of categories—not just species but genera, families, orders, whole kingdoms—based on which anatomical characters they share and which they don't.

All vertebrate animals have backbones. Among vertebrates, birds have feathers, whereas reptiles have scales. Mammals have fur and mammary glands, not feathers or scales. Among mammals, some have pouches in which they nurse their tiny young. Among these species, the marsupials, some have huge rear legs and strong tails by which they go hopping across miles of arid outback; we call them kangaroos. Bring in modern microscopic and molecular evidence, and you can trace the similarities still further back. All plants and fungi, as well as animals, have nuclei within their cells. All living organisms contain DNA and RNA (except some viruses with RNA only), two related forms of information-coding molecules.

Such a pattern of tiered resemblances—groups of similar species nested within broader groupings, and all descending from a single source—isn't naturally present among other collections of items. You won't find anything equivalent if you try to categorize rocks, or musical instruments, or jewelry. Why not? Because rock types and styles of jewelry don't reflect unbroken descent from common ancestors. Biological diversity does. The number of shared characteristics between any one species and another indicates how recently those two species have diverged from a shared lineage.

That insight gave new meaning to the task of taxonomic classification, which had been founded in its modern form back in 1735 by the Swedish naturalist Carolus Linnaeus. Linnaeus showed how species could be systematically classified, according to their shared similarities, but he worked from creationist assumptions that offered no material explanation for the nested pattern he found. In the early and middle nineteenth century, morphologists such as George Cuvier and Étienne Geoffroy Saint-Hilaire in France and Richard Owen in England improved classification with their meticulous studies of internal as well as external anatomies, and tried to make sense of what the ultimate source of these patterned similarities could be. Not even Owen, a contemporary and onetime friend of Darwin's (later in life they had a bitter falling-out), took the full step to an evolutionary vision before *The Origin of Species* was published. Owen made a major contribution, though, by advancing the concept of homologues—that is, superficially different but fundamentally similar versions of a single organ or trait, shared by dissimilar species.

For instance, the five-digit skeletal structure of the vertebrate hand appears not just in humans and apes and raccoons and bears but also, variously modified, in cats and bats and porpoises and lizards and turtles. The paired bones of our lower leg, the tibia and the fibula, are also represented by homologous bones in other mammals and in reptiles, and even in the long-extinct bird-reptile *Archaeopteryx*. What's the reason behind such varied recurrence of a few basic designs? Darwin, with a nod to Owen's "most interesting work," supplied the answer: common descent, as shaped by natural selection, modifying the inherited basics for different circumstances.

Vestigial characteristics are still another form of morphological evidence, illuminating to contemplate because they show that the living world is full of small, tolerable imperfections. Why do male mammals (including human males) have nipples? Why do some snakes (notably boa constrictors) carry the rudiments of a pelvis and tiny legs buried inside their sleek profiles? Why do certain species of flightless beetle have wings, sealed beneath wing covers that never open? Darwin raised all these questions, and answered them, in *The Origin of Species*. Vestigial structures stand as remnants of the evolutionary history of a lineage.

Today the same four branches of biological science from which Darwin drew—biogeography, paleontology, embryology, morphology—embrace an ever growing body of supporting data. In addition to those categories we now have others: population genetics, biochemistry, molecular biology, and, most recently, the whizbang field of machine-driven genetic sequencing known as genomics. These new forms of knowledge overlap one another seamlessly and intersect with the older forms, strengthening the whole edifice, contributing further to the certainty that Darwin was right.

He was right about evolution, that is. He wasn't right about *everything*. Being a restless explainer, Darwin floated a number of theoretical notions during his long working life, some of which were mistaken and illusory. He was wrong about what causes variation within a species. He was wrong about a famous geologic mystery, the parallel

shelves along a Scottish valley called Glen Roy. Most notably, his theory of inheritance—which he labeled pangenesis and cherished
despite its poor reception among his biologist colleagues—turned out
to be dead wrong. Fortunately for Darwin, the correctness of his most
famous good idea stood independent of that particular bad idea. Evolution by natural selection represented Darwin at his best—which is
to say, scientific observation and careful thinking at its best.

DOUGLAS FUTUYMA is a highly respected evolutionary biologist,
author of textbooks as well as influential research papers. His office,
at the University of Michigan, is a long narrow room in the natural
sciences building, well stocked with journals and books, including
volumes about the conflict between creationism and evolution. I
arrived carrying a well-thumbed copy of his own book on that subject, *Science on Trial: The Case for Evolution.* Killing time in the corridor before our appointment, I noticed a blue flyer on a departmental
bulletin board, seeming oddly placed there amid the announcements
of career opportunities for graduate students. "CREATION VS. EVOLU-
TION," it said. "A series of messages challenging popular thought with
Biblical truth and scientific evidences." A traveling lecturer from
something called the Origins Research Association would deliver
these messages at a local Baptist church. Beside the lecturer's photo
was a drawing of a dinosaur. "Free pizza following the evening service," said a small line at the bottom. Dinosaurs, biblical truth, and
pizza: something for everybody.

In response to my questions about evidence, Dr. Futuyma
moved quickly through the traditional categories—paleontology,
biogeography—and talked mostly about modern genetics. He pulled
out his heavily marked copy of the journal *Nature* for February 15,
2001, a historic issue, fat with articles reporting and analyzing the
results of the Human Genome Project. Beside it he slapped down a
more recent issue of *Nature,* this one devoted to the sequenced
genome of the house mouse, *Mus musculus.* The headline of the lead

editorial announced: "Human Biology by Proxy." The mouse genome effort, according to *Nature*'s editors, had revealed "about thirty thousand genes, with 99 percent having direct counterparts in humans."

The resemblance between our thirty thousand human genes and those thirty thousand mousy counterparts, Futuyma explained, represents another form of homology, like the resemblance between a five-fingered hand and a five-toed paw. Such genetic homology is what gives meaning to biomedical research using mice and other animals, including chimpanzees, which (to their sad misfortune) are our closest living relatives.

No aspect of biomedical research seems more urgent today than the study of microbial diseases. And the dynamics of those microbes within human bodies, within human populations, can only be understood in terms of evolution.

Nightmarish illnesses caused by microbes include both the infectious sort (AIDS, Ebola, SARS) that spread directly from person to person and the sort (malaria, West Nile fever) delivered to us by biting insects or other intermediaries. The capacity for quick change among disease-causing microbes is what makes them so dangerous to large numbers of people and so difficult and expensive to treat. They leap from wildlife or domestic animals into humans, adapting to new circumstances as they go. Their inherent variability allows them to find new ways of evading and defeating human immune systems. By natural selection they acquire resistance to drugs that should kill them. They evolve. There's no better or more immediate evidence supporting the Darwinian theory than this process of forced transformation among our inimical germs.

Take the common bacterium *Staphylococcus aureus,* which lurks in hospitals and causes serious infections, especially among surgery patients. Penicillin, becoming available in 1943, proved almost miraculously effective in fighting staphylococcus infections. Its deployment marked a new phase in the old war between humans and disease microbes, a phase in which humans invent new killer drugs and microbes find new ways to be unkillable. The supreme potency of penicillin didn't last long. The first resistant strains of *Staphylococcus aureus* were reported in 1947. A newer staph-killing drug, methicillin, came into use during the 1960s, but methicillin-resistant strains appeared soon, and by the 1980s those strains were widespread. Vancomycin became the next great weapon against staph, and the first vancomycin-resistant strain emerged in 2002. These antibiotic-resistant strains represent an evolutionary series, not much different in principle from the fossil series tracing horse evolution from *Hyracotherium* to *Equus*. They make evolution a very practical problem by adding expense, as well as misery and danger, to the challenge of coping with staph.

The biologist Stephen Palumbi has calculated the cost of treating penicillin-resistant and methicillin-resistant staph infections, just in the United States, at $30 billion a year. "Antibiotics exert a powerful evolutionary force," he wrote last year, "driving infectious bacteria to evolve powerful defenses against all but the most recently invented drugs." As reflected in their DNA, which uses the same genetic code found in humans and horses and hagfish and honeysuckle, bacteria are part of the continuum of life, all shaped and diversified by evolutionary forces.

Even viruses belong to that continuum. Some viruses evolve quickly, some slowly. Among the fastest is HIV, because its method of replicating itself involves a high rate of mutation, and those mutations allow the virus to assume new forms. After just a few years of infection and drug treatment, each HIV patient carries a unique version of the virus. Isolation within one infected person, plus differing conditions and the struggle to survive, forces each version of HIV to evolve independently. It's nothing but a speeded up and microscopic case of what Darwin saw in the Galápagos—except that each human body is an island, and the newly evolved forms aren't so charming as finches or mockingbirds.

Understanding how quickly HIV acquires resistance to antiviral drugs, such as AZT, has been crucial to improving treatment by way of multiple-drug cocktails. "This approach has reduced deaths due to HIV by severalfold since 1996," according to Palumbi, "and it has greatly slowed the evolution of this disease within patients."

Insects and weeds acquire resistance to our insecticides and herbicides through the same process. As we humans try to poison them, evolution by natural selection transforms the population of a mosquito or thistle into a new sort of creature, less vulnerable to that particular poison. So we invent another poison, then another. It's a futile effort. Even DDT, with its ferocious and long-lasting effects throughout ecosystems, produced resistant house flies within a decade of its discovery in 1939. By 1990 more than 500 species (including 114 kinds of mosquitoes) had acquired resistance to at least one pesticide. Based on these undesired results, Stephen Palumbi has commented glumly, "humans may be the world's dominant evolutionary force."

Among most forms of living creatures, evolution proceeds slowly—too slowly to be observed by a single scientist within a research lifetime. But science functions by inference, not just by direct observation, and the inferential sorts of evidence such as paleontology and biogeography are no less cogent simply because they're indirect. Still, skeptics of evolutionary theory ask: Can we see evolution in action? Can it be observed in the wild? Can it be measured in the laboratory?

The answer is yes. Peter and Rosemary Grant, two British-born researchers who have spent decades where Charles Darwin spent weeks, have captured a glimpse of evolution with their long-term studies of beak size among Galápagos finches. William R. Rice and George W. Salt achieved something similar in their lab, through an experiment involving thirty-five generations of the fruit fly *Drosophila melanogaster*. Richard E. Lenski and his colleagues at Michigan State University have done it too, tracking twenty thousand generations of evolution in the bacterium *Escherichia coli*. Such field studies and lab experiments document anagenesis—that is, slow evolutionary change within a single, unsplit lineage. With patience it can be seen, like the movement of a minute hand on a clock.

Speciation, when a lineage splits into two species, is the other major phase of evolutionary change, making possible the divergence between lineages about which Darwin wrote. It's rarer and more elusive even than anagenesis. Many individual mutations must accumulate (in most cases, anyway, with certain exceptions among plants) before two populations become irrevocably separated. The process is spread across thousands of generations, yet it may finish abruptly—like a door going *slam!*—when the last critical changes occur. Therefore it's much harder to witness. Despite the difficulties, Rice and Salt seem to have recorded a speciation event, or very nearly so, in their extended experiment on fruit flies. From a small stock of mated females they eventually produced two distinct fly populations adapted to different habitat conditions, which the researchers judged "incipient species."

AFTER MY VISIT with Douglas Futuyma in Ann Arbor, I spent two hours at the university museum there with Philip D. Gingerich, a paleontologist well-known for his work on the ancestry of whales. As we talked, Gingerich guided me through an exhibit of ancient cetaceans on the museum's second floor. Amid weird skeletal shapes that seemed almost chimerical (some hanging overhead, some in glass cases) he pointed out significant features and described the progress of thinking about whale evolution. A burly man with a broad open face and the gentle manner of a scoutmaster, Gingerich combines intellectual passion and solid expertise with one other trait that's valuable in a scientist: a willingness to admit when he's wrong.

Since the late 1970s Gingerich has collected fossil specimens of early whales from remote digs in Egypt and Pakistan. Working with Pakistani colleagues, he discovered *Pakicetus,* a terrestrial mammal dating from fifty million years ago, whose ear bones reflect its membership in the whale lineage but whose skull looks almost doglike. A former student of Gingerich's, Hans Thewissen, found a slightly more recent form with webbed feet, legs suitable for either walking or swimming, and a long toothy snout. Thewissen called it *Ambulocetus natans,* or the "walking-and-swimming whale." Gingerich and his team turned up several more, including *Rodhocetus balochistanensis,* which was fully a sea creature, its legs more like flippers, its nostrils shifted backward on the snout, halfway to the blowhole position on a modern whale. The sequence of known forms was becoming more and more complete. And all along, Gingerich told me, he leaned toward believing that whales had descended from a group of carnivorous Eocene mammals known as mesonychids, with cheek teeth useful for chewing meat and bone. Just a bit more evidence, he thought, would confirm that relationship. By the end of the 1990s most paleontologists agreed.

Meanwhile, molecular biologists had explored the same question and arrived at a different answer. No, the match to those Eocene carnivores might be close, but not close enough. DNA hybridization and other tests suggested that whales had descended from artiodactyls (that is, even-toed herbivores, such as antelopes and hippos), not from meat-eating mesonychids.

In the year 2000 Gingerich chose a new field site in Pakistan, where one of his students found a single piece of fossil that changed the prevailing view in paleontology. It was half of a pulley-shaped anklebone, known as an astragalus, belonging to another new species of whale. A Pakistani colleague found the fragment's other half. When Gingerich fitted the two pieces together, he had a moment of humbling recognition: The molecular biologists were right. Here was an anklebone, from a four-legged whale dating back forty-seven million years, that closely resembled the homologous anklebone in an artiodactyl. Suddenly he realized how closely whales are related to antelopes.

This is how science is supposed to work. Ideas come and go, but the fittest survive. Downstairs in his office Phil Gingerich opened a specimen drawer, showing me some of the actual fossils from which the display skeletons upstairs were modeled. He put a small lump of petrified bone, no larger than a lug nut, into my hand. It was the famous astragalus, from the species he had eventually named *Artiocetus clavis*. It felt solid and heavy as truth.

Seeing me to the door, Gingerich volunteered something personal:

"I grew up in a conservative church in the Midwest and was not taught anything about evolution. The subject was clearly skirted. That helps me understand the people who are skeptical about it. Because I come from that tradition myself." He shares the same skeptical instinct. Tell him that there's an ancestral connection between land animals and whales, and his reaction is: Fine, maybe, but show me the intermediate stages. Like Charles Darwin, the onetime divinity student, who joined that round-the-world voyage aboard the *Beagle* instead of becoming a country parson, and whose grand view of life on Earth was shaped by close attention to small facts, Phil Gingerich is a reverent empiricist. He's not satisfied until he sees solid data. That's what excites him so much about pulling whale fossils out of the ground. In thirty years he has seen enough to be satisfied. For him, Gingerich said, it's "a spiritual experience."

"The evidence is there," he added. "It's buried in the rocks of ages."

On the Origins of the Mind

FROM *COMMENTARY*

It's all in our genes—even what's in our heads. At least, that's what those seeking an evolutionary explanation of human behavior might say. In this provocative essay on evolutionary psychology and the nature of the mind, the mathematician David Berlinski shows how much we don't know in both subjects.

It's all scientific stuff; it's been proved.
—TOM BUCHANAN in *The Great Gatsby*

At some time in the history of the universe, there were no human minds, and at some time later, there were. Within the blink of a cosmic eye, a universe in which all was chaos and void came to include hunches, beliefs, sentiments, raw sensations, pains, emotions, wishes, ideas, images, inferences, the feel of rubber, Schadenfreude, and the taste of banana ice cream.

A sense of surprise is surely in order. How did *that* get *here*?

If the origin of the human mind is mysterious, so too is its nature. There are, Descartes argued, two substances in the universe, one physical and the other mental.

To many contemporary philosophers, this has seemed rather an embarrassment of riches. But no sooner have they ejected mental substances from their analyses than mental properties pop up to take their place, and if not mental properties then mental functions. As a conceptual category, the mental is apparently unwilling to remain expunged.

And no wonder. Although I may be struck by a thought, or moved by a memory, or distracted by a craving, these familiar descriptions suggest an effect with no obvious physical cause. Thoughts, memories, cravings—they are what? Crossing space and time effortlessly, the human mind deliberates, reckons, assesses, and totes things up; it reacts, registers, reflects, and responds. In some cases, like inattention or carelessness, it invites censure by doing nothing at all or doing something in the wrong way; in other cases, like vision, it acts unhesitatingly and without reflection; and in still other cases, the human mind manages both to slip itself into and stay aloof from the great causal stream that makes the real world boom, so that when *it* gives the go-ahead, what *I* do is, as Thomas Aquinas observed, "inclined but not compelled."

These are not properties commonly found in the physical world. They are, in fact, not found at all.

And yet, the impression remains widespread that whoever is responsible for figuring out the world's deep things seems to have figured out the human mind as well. Commenting on negative advertising in political campaigns, Kathleen Hall Jamieson, the director of the Annenberg Public Policy Center at the University of Pennsylvania, remarked that "there appears to be something hard-wired into humans that gives special attention to negative information." There followed what is by now a characteristic note: "I think it's evolutionary biology."

Negative campaign advertisements are the least of it. There is, in addition, war and male aggression, the human sensitivity to beauty, gossip, a preference for suburban landscapes, love, altruism, marriage, jealousy, adultery, road rage, religious belief, fear of snakes, disgust, night sweats, infanticide, and the fact that parents are often fond of their children. The idea that human behavior is "the product of evolution," as the *Washington Post* puts the matter, is now more than a theory: it is a popular conviction.

It is a conviction that reflects a modest consensus of opinion among otherwise disputatious philosophers and psychologists: Steven Pinker, Daniel Dennett, David Buss, Henry Plotkin, Leda Cosmides, John Tooby, Peter Gärdenfors, Gary Marcus. The consensus is constructed, as such things often are, on the basis of a great hope and a handful of similes. The great hope is that the human mind will in the end find an unobtrusive place in the larger world in which purely material causes chase purely material effects throughout the endless night. The similes are, in turn, designed to promote the hope.

Three similes are at work, each more encompassing than the one before. They give a natural division of labor to what is now called evolutionary psychology.

First, the human mind is *like* a computer in the way that it works. And it is just because the mind *is* like a computer that the computer comprises a model of the mind. "My central thesis," the cognitive psychologist H. A. Simon has written, is that "conventional computers can be, and have been, programmed to represent symbol structures and carry out processes on those structures that parallel, step by step, the way the human brain does it."

Second, the individual human mind is *like* the individual human kidney, or any other organ of the body, in the way that it is created anew in every human being. "Information," Gary Marcus writes, "copied into the nucleus of every newly formed cell, guides the gradual but powerful process of successive approximation that shapes each of the body's organs." This is no less true of the "organ of thought and language" than of the organs of excretion and elimination.

Third, the universal human mind—the expression in matter of human nature—is *like* any other complicated biological artifact in the way that it arose in the human species by means of random variation and natural selection. These forces, as Steven Pinker argues, comprise "the only explanation we have of how complex life *can* evolve. . . ."

Taken together, these similes do succeed wonderfully in suggesting a coherent narrative. The ultimate origins of the human mind may be found in the property of irritability that is an aspect of living tissue itself. There is a primordial twitch, one that has been lost in time but not in memory; various descendant twitches then enlarged themselves, becoming, among the primates at least, sophisticated organs of perception, cognition, and computation. The great Era of Evolutionary Adaptation arrived in the late Paleolithic, a veritable genetic Renaissance in which the contingencies of life created, in the words of the evolutionary psychologist Leda Cosmides, "programs that [were] well-engineered for solving problems such as hunting, foraging for plant foods, courting mates, cooperating with kin, forming coalitions for mutual defense, avoiding predators, and the like." There followed the long Era in Which Nothing Happened, the modern human mind retaining in its structure and programs the mark of the time that human beings spent in the savannah or on the forest floor, hunting, gathering, and reproducing with Darwinian gusto.

Three quite separate scientific theories do much to justify this grand narrative and the three similes that support it. In the first instance, computer science; in the second, theories of biological development; in the third, Darwin's theory of evolution. At times, indeed, it must seem that only the width of a cigarette paper separates evolutionary psychology from the power and the glory of the physical sciences themselves.

The Model for What Science Should Be

IF THE CLAIMS of evolutionary psychology are ambitious, the standard against which they should be assessed must be mature, reasonable, and persuasive. If nothing else, that standard must reflect principles that have worked to brilliant success in the physical sciences themselves. This is more than a gesture of respect; it is simple common sense.

In stressing the importance of their subject, the mathematicians J. H. Hubbard and B. H. West begin their textbook on differential equations by observing that "historically, Newton's spectacular success in describing mechanics by differential equations was a *model for what science should be*" (emphasis added). Hubbard and West then add what is in any case obvious: that "all basic physical laws are stated as differential equations, whether it be Maxwell's equations for electrodynamics, Schrödinger's equation for quantum mechanics, or Einstein's equations for general relativity."

Equations do lie close to the mathematician's heart, and differential equations closer than most. On one side of such an equation, there is a variable denoting an unknown mathematical function; on the other, a description of the rate at which that unknown function is changing at every last moment down to the infinitesimal. Within the physical sciences, such changes express the forces of nature, the moon perpetually falling because perpetually accelerated by the universal force of gravitation. The mathematician's task is to determine the overall, or global, identity of the unknown function from its local rate of change.

In describing the world by means of a differential equation, the mind thus moves from what is local to what is global. It follows that the "model for what science should be" involves an interdiction against action at a distance. "One object," the Russian mathematician Mikhael Gromov observes, "cannot influence another one removed from it without involving local agents located one next to another and making a continuous chain joining the two objects." As for what happens when the interdiction lapses, Gromov, following the French mathematician René Thom, refers to the result as *magic*. This contrast between a disciplined, differential description of a natural process and an essentially magical description is a useful way of describing a fundamental disjunction in thought.

A differential equation, it is important to stress, offers only a general prescription for change. The distance covered by a falling object is a matter of how fast it has been going and how long it has been going fast; this, an equation describes. But how *far* an object has gone depends on how high it was when it began falling, and this the underlying equation does not specify and so cannot describe. The solutions to a differential equation answer the question, how is the process changing? The data themselves answer a quite different question: how or where does the process *start?* Such specifications comprise the initial conditions of a differential equation, and represent the intrusion into the mathematical world of circumstances beyond the mathematical.

It is this that in 1902 suggested to the French mathematician Jacques Hadamard the idea of a "well-posed problem" in analysis. For a differential equation to be physically useful, Hadamard argued, it must meet three requirements. Solutions must in the first place exist. They must be unique. And they must in some reasonable sense be stable, the solutions varying continuously as the initial conditions themselves change.

With these requirements met, a well-posed differential equation achieves a coordination among continuous quantities that is determined for every last crack and crevice in the manifold of time. And is this the standard I am urging on evolutionary psychology? Yes, absolutely.

Nothing but the best.

That the Human Mind Is Like a Digital Computer

ALTHOUGH EVOLUTIONARY PSYCHOLOGISTS have embraced the computational theory of mind, it is not entirely a new theory; it has been entertained, if not embraced, in other places and at other times. Gottfried Leibniz wrote of universal computing machines in the seventeenth century, and only the limitations of seventeenth-century technology prevented him from toppling into the twenty-first. As it was, he did manage to construct a multipurpose calculator, which, he claimed, could perform the four elementary operations of addition, subtraction, division, and multiplication. But when he demonstrated the device to members of the Royal Society in London, someone in the wings noticed that he was carrying numbers by hand.

I do not know whether this story is true, but it has a very queer power, and in a discussion dominated by any number of similes it constitutes a rhetorical figure—shaped as a warning—all its own.

In 1936 the British logician Alan Turing published the first of his papers on computability. Using nothing more than ink, paper, and the resources of mathematical logic, Turing managed to create an imaginary machine capable of incarnating a very smooth, very suave imitation of the human mind.

Known now as a Turing machine, the device has at its disposal a tape divided into squares and a reading head mounted over the tape. It has, as well, a finite number of physical symbols, most commonly zeros and ones. The reading head may occupy one of a finite number of distinct physical states. And thereafter the repertoire of its action is extremely limited. A Turing machine can, in the first place, recognize symbols, one square at a time. It can, in the second place, print symbols or erase them from the square it is scanning. And it can, in the third place, change its internal state, and move to the left or to the right of the square it is scanning, one square at a time.

There is no fourth place. A Turing machine can do nothing else. In fact, considered simply as a mechanism, a Turing machine can do nothing whatsoever, the thing existing in that peculiar world—my own, and I suspect others' as well—in which everything is possible but nothing gets done.

A Turing machine gains its powers of imitation only when, by means of a program, or an algorithm, it is told what to do. The requisite instructions consist of a finite series of commands, written in a stylized vocabulary precisely calibrated to take advantage of those operations that a Turing machine can perform. What gives to the program its air of cool command is the fact that its symbols function in a double sense. They are symbols by virtue of their *meaning,* and so reflect the intentions of the human mind that has created them; but they are *causes* by virtue of their structure, and so enter into the rhythms of the real world. Like the word "bark," which both expresses a human command and sets a dog to barking, the symbols do double duty.

Although imaginary at its inception, a Turing machine brilliantly anticipated its own realization in matter. Through a process of intellectual parthenogenesis, Turing's ideas gave rise to the modern digital computer. And once the sheer physical palpability of the computer was recognized—there it is, as real as the rocks, the rifts, and the rills of the physical sciences—there was nothing to stand in the way of the first controlling simile of evolutionary psychology: that the human mind is itself a computer, one embodied in the human brain.

The promotion of the computer from an imaginary to a physical object serves the additional purpose of restoring it to the world that can be understood in terms of the "model for what science should be." As a physical device, nothing more than a collection of electronic circuits, the digital computer can be represented entirely by Clerk Maxwell's theory of the electromagnetic field, with the distinction between a Turing machine and its program duplicated in the distinction between a differential equation and its initial conditions. We are returned to the continuous and infinite world studied by mathematical physics, the world in which differential equations track the evolution of material objects moving through time in response to the eternal forces of nature itself.

The intellectual maneuvers that I have recounted serve to make the computer an irresistibly compelling object. But they serve, as well, to displace attention from the human mind. The effect is to endow the simile that the human mind is like a computer with a plausibility it might not otherwise enjoy.

A certain "power to alter things," Albertus Magnus observed, "indwells in the human soul." The *existence* of this power is hardly in doubt. It is evident in every human act in which the mind imposes itself on nature by taking material objects from their accustomed place and rearranging them; and it is evident again whenever a human being interacts with a machine. Writing with characteristic concision in the *Principia*, Isaac Newton observed that "the power and use of machines consist only in this, that by diminishing the velocity *we* may augment the force, and the contrary" (emphasis added). Although Newton's analysis was restricted to mechanical forces (he knew no others), his point is nonetheless general. A machine is a material object, a *thing*, and as such, its capacity to do work is determined by the forces governing its behavior and by its initial conditions.

Those initial conditions must themselves be explained, and in the nature of things they cannot be explained by the very device that *they* serve to explain. This is precisely the problem that Newton faced in the *Principia*. The magnificent "system of the world" that he devised explained why the orbits of the planets around the Sun must be represented by a conic section; but Newton was unable to account for the initial conditions that he had himself imposed on his system. Facing an imponderable, he appealed to divine intervention. It was not until Pierre Simon Laplace introduced his nebular hypothesis in 1796 that some form of agency was removed from Newtonian mechanics.

This same pattern, along with the problem it suggests, recurs whenever machines are at issue, and it returns with a vengeance whenever computers are invoked as explanations for the human mind. A computer is simply an electromechanical device, and this is precisely why it is useful as a model of the human brain. By setting its initial conditions, a computer's program allows the machine to do work in the real world. But the normal physical processes by which a computer works are often obscured by their unfamiliarity—who among us *really* understands what a computer is and how it works? No doubt, this is why the thesis that the mind is like a computer resonates with a certain intellectual grandeur.

An abacus conveys no comparable air of mystery. It is a trifle. Made of wood, it consists of a number of wires suspended in a frame and a finite number of beads strung along the wires. Nevertheless, an idealized abacus has precisely the power of a Turing machine, and so both the abacus and the Turing machine serve as models for a working digital computer. By parity of reasoning, they also both serve as models for the human mind.

Yet the thesis that the human mind is like an abacus seems distinctly less plausible than the thesis that the human mind is like a computer, and for obvious reasons. It is precisely when things have been reduced to their essentials that the interaction between a human being and a simple machine emerges clearly. That interaction is naked, a human agent handling an abacus with the same directness of touch that he might employ in handling a lever, a pulley, or an inclined plane. The force that human beings bring to bear on simple machines is muscular and so derived from the chemistry of the human body, the causes ultimately emptying out into the great ocean of physical interactions whose energy binds and loosens the world's large molecules. But what we need to know in the example of the abacus is not the nature of the forces controlling its behavior but the circumstances by which those forces come into play.

No chain of causes known to date accommodates the inconvenient fact that, by setting the initial conditions of a simple machine, a human agent brings about a novel, an unexpected, an entirely idiosyncratic distribution of matter. Every mechanical artifact represents what the anthropologist Mary Douglas calls "matter out of place." The problem that Newton faced but could not solve in the *Principia* returns when an attempt is made to provide a description of the simplest of human acts, the trivial tap or touch that sets a polished wooden bead spinning down a wire. Tracing the causal chain backward leads only to a wilderness of causes, each of them displacing material objects from their proper settings, so that in the end the mystery is simply shoveled back until the point is reached when it can be safely ignored.

A chain of physical causes is thus not obviously useful in explaining how a human agent exhibits the capacity to "alter things." But neither does it help to invoke, as some have done, the hypothesis that another abacus is needed to fix the initial conditions of the first. If each abacus requires yet another abacus in turn, the road lies open to the madness of an infinite regress, a point observed more than seventy years ago by the logicians Kurt Gödel and Alfred Tarski in their epochal papers on incompleteness.

If we are able to explain how the human mind works neither in terms of a series of physical causes nor in terms of a series of infinitely receding mechanical devices, what then is left? There is the ordinary, very rich, infinitely moving account of mental life that without hesitation we apply to ourselves. It is an account frankly magical in its nature. The human mind registers, reacts, and responds; it forms intentions, conceives problems, and then, as Aristotle dryly noted, it *acts*. In analyzing action, we are able to say only, as Albertus Magnus said, that a certain power to alter things inheres in the human soul.

A simile that for its persuasiveness depends on the very process it is intended to explain cannot be counted a great success.

That the Human Mind Is Like Any Other Organ of the Body

IF THE COMPUTATIONAL ACCOUNT of the human mind cannot be brought under the control of the "model for what science should be," what of the thesis that the human mind can be comprehended by reference to the laws of biological development? Here we come to the second simile of evolutionary psychology.

"As the ruler of the soul," Ptolemy wrote in the *Tetrabiblos*, "Saturn has the power to make men sordid, petty, mean-spirited, indifferent, mean-minded, malignant, cowardly, diffident, evil-speaking, solitary, tearful, shameless, superstitious, fond of toil, unfeeling, devisors of plots against their friends, gloomy, taking no care of their body." We know the type; there is no need to drown the point in words. Some men are just rotten.

The analysis that Ptolemy offers in defense of his thesis is anything but crude. "The chronological starting point of human nativities," he writes, "is naturally the very time of conception, for to the seed is given once and for all the very qualities that will mark the adult and that are expressed in growth." It is Saturn's position that affects the seed, and the seed thereafter that affects the man.

Ptolemy's sophistication notwithstanding, no one today is much minded to study the *Tetrabiblos* as a guide to human psychology. Even if a convincing correlation could be established between the position of the planets and the onset of human rottenness, persuading us that we have identified some remote cause in nature for some human effect, that cause would quite obviously violate the interdiction against action at a distance. Ptolemy himself was sensitive to the distinction between astrological knowledge and real knowledge. In trying to construct a continuous chain between the position of the planets and the advent of human rottenness, he was at as great a loss as we are. It is for this reason that the word he employs to describe the way in which heavenly objects evoke their effects is *influence;* it is a word that does not appear, and is not needed, in the *Almagest,* Ptolemy's great treatise on astronomy.

More than two thousand years have gone by since Ptolemy composed the *Tetrabiblos.* The stars have withdrawn themselves; their role in human affairs has been assigned to other objects. Under views accepted by every evolutionary psychologist, the source of human rottenness may be found either in the environment or within the human genome.

The first of these, the environment, has been the perpetual Plaintiff of Record in *Nurture* v. *Nature et al.* But for our purposes it may now be dismissed from further consideration. If some men are made bad, then they are not born that way; and if they are not born that way, an explanation of human rottenness cannot be expressed in evolutionary terms.

The question at hand is thus whether the path initiated by the human genome in development can be understood in terms of "the model for what science should be." A dynamical system is plainly at work, one that transforms what Ptolemy called "the seed" into a fully formed human being in nine months, and then into an accomplished car thief in less than twenty years. What evolutionary psychology requires is a demonstration that this process may itself be brought under control of a description meeting the standard that "one object cannot influence another one removed from it without involving local agents located one next to another and making a continuous chain joining the two objects."

Well, can it?

"OUR BASIC PARADIGM," Benjamin Levin writes in his textbook on genetics, "is that genes encode proteins, which in turn are responsible for the synthesis of other structures." Levin is a careful and a conscientious writer. By "other structures" he means only the nucleic acids. But his "basic paradigm" is now a part of a great cultural myth, and by "other structures" most evolutionary psychologists mean *all* of the structures that are made from the proteins, most notably the human brain.

The myth begins solidly enough—with the large bio-molecules that make up the human genome. The analysis of the precise, unfathomably delicate steps that take place as the genome undertakes various biochemical activities has been among the glories of modern science. Unfortunately, however, the chain of causes that begins in the human genome gutters out inconclusively long before the chain can be continued to the human brain, let alone the human mind. Consider in this regard the following sequence of representative quotations in which tight causal connections are progressively displaced in favor of an ever more extravagant series of metaphors:

(1) *Quantum chemistry:* "For a molecule, it is reasonable to split the kinetic energy into two summations—one over the electrons, and one over the nuclei."

(2) *Biochemistry:* "Initiation of prokaryotic translation requires a tRNA bearing N-formyl methionne, as well as three initiation factors (IF1, 2, 3), a 30S ribosomal subunit GTP," etc.

(3) *Molecular biology:* "Once the protein binds one site, it *reaches* the other by *crawling* along the DNA, thus *preserving* its *knowledge* of the orientation of the first site" (emphasis added).

(4) *Embryology:* "In the embryo, cells divide, *migrate, die, stick to each other, send out* processes, and *form* synapses" (emphasis added).

(5) and (6) *Developmental genetics:* "But genes are simply regulatory elements, molecules that *arrange* their surrounding environments into an *organism*" (emphasis added).

"Genes *prescribe* epigenetic *rules,* which are the neural *pathways* and *regularities* in *cognitive development* by which the individual *mind assembles* itself" (emphasis added).

(7) *Developmental biology:* "The *pattern* of neural connections (synapses) *enables* the human cortex to *function* as the *center* for *learning, reasoning,* and *memory,* to *develop* the *capacity* for *symbolic expression,* and to *produce voluntary responses* to interpreted stimuli" (emphasis added).

(8) and (9) *Evolutionary psychology:* "Genes, of course, do *influence* human development" (emphasis added).

"[Genes] *created* us, body and mind" (emphasis added).

Now the very sober (1) and (2) are clearly a part of "the model for what science should be." By the time we come to (3), however, very large molecular chains have acquired powers of agency: they are busy reaching, crawling, and knowing; it is by no means clear that these metaphors may be eliminated in favor of a biochemical description. Much the same is true of (4). In (5) and (6), a connection is suggested between genes, on the one hand, and organisms, on the other, but the chain of causes and their effects has become very long, the crucial connections now entirely expressed in language that simply disguises profound gaps in our understanding.

In (7) the physical connection between morphology and the mind is reduced to wind, while (8) defiantly resurrects "influence," Ptolemy's original term of choice. It is the altogether exuberant (9)— the quotation is from Richard Dawkins—that finally drowns out any last faint signal from the facts.

These literary exercises suggest that the longer the chain of causes, the weaker the links between and among them. Whether this represents nothing more than the fact that our knowledge is incomplete, or whether it points to a conceptual deficiency that we have no way of expressing, let alone addressing—these are matters that we cannot now judge.

Curiously enough, it has been evolutionary psychologists themselves who are most willing to give up in practice what they do not have in theory. For were that missing theory to exist, it would cancel—it would *annihilate*—any last lingering claim we might make on behalf of human freedom. The physical sciences, after all, do not simply trifle with determinism: it is the heart and soul of their method. Were boron salts at liberty to discard their identity, the claims of inorganic chemistry would seem considerably less pertinent than they do.

Thus, when Steven Pinker writes that "nature does not dictate what we should accept or how we should live our lives," he is expressing a hope entirely at odds with his professional commitments. If ordinary men and women are, like the professor himself, perfectly free to tell their genes "to go jump in the lake," why then pay the slightest attention to evolutionary psychology—why pay the slightest attention to Pinker?

Irony aside, a pattern is at work here. Where (in the first simile) computational accounts of the mind are clear enough to be encompassed by the model for what science should be, they are incomplete—radically so. They embody what they should explain. Where (in the second simile) biochemical and quantum chemical accounts of development are similarly clear and compelling, they extend no farther than a few large molecules. They defer what they cannot explain. In both cases, something remains unexplained.

This is a disappointing but perhaps not unexpected conclusion. We are talking, after all, about the human mind.

That the Human Mind Is Like Any Other Biological Artifact

EVOLUTIONARY PSYCHOLOGISTS believe that the only force in nature adequate to the generation of biological complexity is natural selection. It is an axiom of their faith. But although natural selection is often described as a force, it is certainly not a force of *nature*. There are four such forces in all: gravitational, electromagnetic, and the strong and weak forces. Natural selection is not one of them. It appears, for the most part, as a free-floating form of agency, one whose identity can only be determined by field studies among living creatures—the ant, the field mouse, and the vole.

But field studies have proved notoriously inconclusive when it comes to natural selection. After three decades spent observing Darwin's finches in the Galapagos, P. R. and B. R. Grant were in the end able to state only that "further continuous long-term studies are needed." It is the conclusion invariably established by evolutionary field studies, and it is the only conclusion established with a high degree of reliability.

The largest story told by evolutionary psychology is therefore anecdotal. Like other such stories, it subordinates itself to the principle that we are what we are because we were what we were. Who could argue otherwise? All too often, however, this principle is itself supported by the counterprinciple that we were what we were because we are what we are, a circle not calculated to engender confidence.

Thus, in tests of preference, Victor Johnson, a bio-psychologist at New Mexico State University, has reported that men throughout the world designate as attractive women with the most feminine faces. Their lips are large and lustrous, their jaws narrow, their eyes wide. On display in every magazine and on every billboard, such faces convey "accented hormonal markers." These are a guide to fertility, and it is the promise of fertility that prompts the enthusiastic male response.

There is no reason to doubt Johnson's claim that on the whole men prefer pretty young women to all the others—the result, I am sure, of research extending over a score of years. It is the connection to fertility that remains puzzling. If male standards of beauty are rooted in the late Paleolithic era, men worldwide should now be looking for stout muscular women with broad backs, sturdy legs, a high threshold to pain, and a welcome eagerness to resume foraging directly after parturition. It has not been widely documented that they do.

In any case, an analysis of human sexual preferences that goes no farther than preferences is an exercise in tiptoeing to the threshold of something important and never peering over. The promise of evolutionary psychology is nothing less than an explanation of the human *mind*. No psychological theory could possibly be considered complete or even interesting that did not ask *why* men exhibit the tastes or undertake the choices they do. When it comes to sexual "preferences," what is involved is the full apparatus of the passions—beliefs, desires, sentiments, wishes, hopes, longings, aching tenderness. To study preferences without invoking the passions is like studying lightning without ever mentioning electricity.

This is one of those instances where evolutionary psychology betrays a queer family resemblance to certain theories in philosophy and psychology that (as we have seen in the case of determinism) evolutionary psychologists are themselves eager to disown. Behaviorism in psychology, as in the work of John Watson and B. F. Skinner, came to grief because human behavior is itself a contested category, and one that lapses into irrelevance once it is enlarged to accommodate the sources of behavior in the mind itself. It may be possible to analyze the mating strategies of the vole, the subject of much current research, by means of a simple assessment of what the vole does: a single genetic switch seems sufficient to persuade an otherwise uxorious male vole to become flamboyantly promiscuous. But human beings, it goes without saying, are not voles, and what *they* do becomes intelligible to them only when it is coordinated with what they are.

Despite the palpably unreliable stories that evolutionary psychologists tell about the past, *is* there, nevertheless, a scientifically reasonable structure that may be invoked to support those stories (as fine bones may support an otherwise frivolous face)?

The underlying tissue that connects the late Paleolithic and the modern era is the gene pool. Changes to that pool reflect a dynamic process in which genes undergo change, duplicate themselves, surge into the future or shuffle off, and by means of all the contingencies of life serve in each generation the purpose of creating yet another generation. This is the province of population genetics, a discipline given a remarkably sophisticated formulation in the 1930s and 1940s by Ronald Fisher, J. B. S. Haldane, and Sewall Wright. Excellent mathematicians, these men were interested in treating evolution as a process expressed by some underlying system of equations. In the 1970s and

1980s, the Japanese population geneticist Motoo Kimura revived and then extended their theories.

Kimura's treatise, *The Neutral Theory of Molecular Evolution* (1983), opens with words that should prove sobering to any evolutionary psychologist: "The neutral theory asserts that the great majority of evolutionary changes at the molecular level, as revealed by comparative studies of protein and DNA sequences, are caused not by Darwinian selection but by random drift of selectively neutral or nearly neutral mutants."

If Darwin's theory is a matter of random variation *and* natural selection, it is natural selection that is demoted on Kimura's view. Random variation is paramount; chance is the driving force. This is carefully qualified: Kimura is writing about "the great majority of evolutionary changes," not all. In addition, he is willing to accept the Darwinian disjunction: either complex adaptations are the result of natural selection or they are the result of nothing at all. But the effect of his work is clear: insofar as evolution is neutral, it is not adaptive, and insofar as it is not adaptive, natural selection plays no role in life.

Like his predecessors, Kimura writes within a particular tradition, one whose confines are fixed by the "model for what science should be." Thus, in trying to describe the fate of a mutant gene, Kimura is led to a differential equation—the Fokker-Planck equation, previously used to model diffusion processes. Although complicated, the equation has a straightforward interpretation. It describes the evolution of a probability distribution, tracking the likelihood over every instant of time that a specific gene will change its representation in a population of genes. Kimura is able to provide an explicit solution for the equation, and thus to treat molecular evolution as a well-posed problem in analysis.

But if the "model for what science should be" is powerful, it is also limited. Stretching it beyond its natural limits often turns out to be an exercise in misapplied force, like a blow delivered to the empty air.

As I have noted several times, the power of a differential equation to govern the flow of time is contingent on some specification of its initial conditions. It is precisely these initial conditions that anecdotal accounts of human evolution cannot supply. We can say of those hunters and gatherers only that they hunted and they gathered, and we can say this only because it seems obvious that there was nothing else for them to do. The gene pool that they embodied cannot be directly recovered.

The question very naturally arises: might that gene pool be recovered from the differential equations of mathematical genetics, much as the original position and momentum of a system of particles moving under the influence of gravitational forces might be recovered from their present position and momentum? This is the question posed by Richard Lewontin.* Writing in a recent issue of the *Annual Review of Genetics,* Lewontin observes that if Kimura's equations carry "a population forward in time from some initial conditions," then what is needed is a second theory, one "that can reverse the deductions of the first theory and infer backward from a particular observed state at present."

Lewontin is correct: this is precisely what is needed. Given the trajectory described by the solution of the Fokker-Planck equation, it *is* certainly possible to track the equation backward, past the middle ages, well past the Roman and then the Sumerian empires, and then into the era of the hunter-gatherers. There is nothing troubling about this. Kimura's equation has an explicit solution, and seeing where it led from is like running a film backward.

But whether, in running this particular film backward, we inevitably channel the temporal stream into a *unique* set of initial conditions is not altogether clear. With questions of this sort, we are in the domain of inverse problems, in which the past is contingent on the present. The solution to an inverse problem, the Russian mathematician Oleg Alifanov remarked, "entails determining unknown causes based on observation of their effects." It is this problem that

---

*I am grateful to Robert Berwick of MIT for calling my attention to this article, and for insisting on its importance.

evolutionary psychology must solve if its engaging stories about the Paleolithic era are to command credibility at the molecular level.

And it is this problem that Lewontin argues cannot be solved in the context of mathematical genetics. "A dynamical theory that predicts the present state generally requires that we know not only the nature and magnitude of the forces that have operated, but also the initial conditions and how long the process has been in operation." This double requirement—*know the forces, specify the initial conditions*—cannot simultaneously be met in going backward from the present. One item of knowledge is needed for the other.

This specific argument may now be enlarged to accommodate the general case. Inverse problems arise in mathematics when the attempt is made to run various mathematical films backward, and they are by now sufficiently well understood so that something may be said about them in a rough-and-ready way. Inverse problems are *not* in general well posed. Observing a pot of boiling liquid, we cannot use the heat equations to determine its identity. Many liquids reach the same boiling point in roughly the same time.

With inverse problems, what is, in fact, lost is the essential sureness and power of the "model for what science should be," and we are returned to a familiar world in which things and data are messy, disorganized, and partial, and in which theories, despite our best intentions, find themselves unable to peep over the hedge of time into the future or the past.

A familiar and by now depressing shape has revealed itself beneath the third and final simile of evolutionary psychology. It succeeds in meeting the demands of "the model for what science should be," but it succeeds in meeting those demands only at an isolated point. The rest is darkness, mystery, and magic.

The Origins of the Human Mind

IF THE CHIEF SIMILES of evolutionary psychology have not improved our understanding of the human mind in any appreciable sense, might we at least say that they have done something toward promoting the field's principal hope, namely, that the mind will in the end take its place as a material object existing in a world of other material objects?

This too is by no means clear. As Leda Cosmides has very sensibly observed, evolutionary psychology is more a research program than a body of specific results. As a program, it rather resembles a weekend athlete forever preparing to embark on a variety of strenuous exercises. In the literature of evolutionary psychology, there is thus no very determined effort to assess any of the classical topics in the philosophy of mind with the aim of doing more than affirming vaguely that some aspect of the mind exists because it may well have been useful. There is, in evolutionary psychology, no account of the emotions beyond the trivial, or of the sentiments, no account of action or intention, no account of the human ability to acquire mathematical or scientific knowledge, no very direct exploration of the mind's power to act at a distance by investing things with meaning—no account, that is, of any of the features of the mind whose existence prompts a question about its origins. In its great hope as in so many other respects, evolutionary psychology has reposed its confidence on the bet that in time these things will be explained. If that is so, all that we on the outside can say is that time will tell.

Yet any essay on evolutionary psychology would be incomplete if it did not acknowledge the moving power of its chief story. For that story, involving as it does our own ancestors, suggests that the human mind that we now occupy had its source in circumstances that, although occluded by time and damaged by distance, are nonetheless familiar.

The time is the distant past. "In Babylonia," the third-century historian Eusebius writes in recounting the lost histories of Berossos the Chaldean, a large number of people "lived without discipline and without order, just like the animals." A frightening monster named Oannes then appeared to the Babylonians after clambering out of the Red Sea. "It had the whole body of a fish, but underneath and attached to the head of the fish there was another head, human, and joined to the tail of the fish, feet, like those of a man, and it had a human voice." The monster "spent his days with men, never eating anything, but teaching men the skills necessary for writing, and for doing mathematics, and for all sorts of knowledge."

Since that time, Eusebius adds regretfully, "nothing further has been discovered."

MARK SOLMS

Freud Returns

FROM *SCIENTIFIC AMERICAN*

With the rise of new therapies and medications, Freudian psycho-analysis has lost its dominance among psychiatrists. But, as the neuro-scientist Mark Solms points out, new developments in the study of the brain are producing results that fit surprisingly well with Freud's theory.

For the first half of the 1900s, the ideas of Sigmund Freud dominated explanations of how the human mind works. His basic proposition was that our motivations remain largely hidden in our unconscious minds. Moreover, they are actively withheld from consciousness by a repressive force. The executive apparatus of the mind (the ego) rejects any unconscious drives (the id) that might prompt behavior that would be incompatible with our civilized conception of ourselves. This repression is necessary because the drives express themselves in unconstrained passions, childish fantasies, and sexual and aggressive urges.

Mental illness, Freud said until his death in 1939, results when repression fails. Phobias, panic attacks, and obsessions are caused by intrusions of the hidden drives into voluntary behavior. The aim of psychotherapy, then, was to trace neurotic symptoms back to their unconscious roots and expose these roots to mature, rational judgment, thereby depriving them of their compulsive power.

As mind and brain research grew more sophisticated from the 1950s onward, however, it became apparent to specialists that the evidence Freud had provided for his theories was rather tenuous. His principal method of investigation was not controlled experimentation but simple observations of patients in clinical settings, interwoven with theoretical inferences. Drug treatments gained ground, and biological approaches to mental illness gradually overshadowed psychoanalysis. Had Freud lived, he might even have welcomed this turn of events. A highly regarded neuroscientist in his day, he frequently made remarks such as "the deficiencies in our description would presumably vanish if we were already in a position to replace the psychological terms by physiological and chemical ones." But Freud did not have the science or technology to know how the brain of a normal or neurotic personality was organized.

By the 1980s the notions of ego and id were considered hopelessly antiquated, even in some psychoanalytical circles. Freud was history. In the new psychology, the updated thinking went, depressed people do not feel so wretched because something has undermined their earliest attachments in infancy—rather their brain chemicals are unbalanced. Psychopharmacology, however, did not deliver an alternative grand theory of personality, emotion, and motivation—a new conception of "what makes us tick." Without this model, neuroscientists focused their work narrowly and left the big picture alone.

Today that picture is coming back into focus, and the surprise is this: it is not unlike the one that Freud outlined a century ago. We are still far from a consensus, but an increasing number of diverse neuroscientists are reaching the same conclusion drawn by Eric R. Kandel of Columbia University, the 2000 Nobel laureate in physiology or medicine: that psychoanalysis is "still the most coherent and intellectually satisfying view of the mind."

Freud is back, and not just in theory. Interdisciplinary work groups uniting the previously divided and often antagonistic fields of neuro-science and psychoanalysis have been formed in almost every major city of the world. These networks, in turn, have come together as the International Neuro-Psychoanalysis Society, which organizes an annual congress and publishes the successful journal *Neuro-Psycho-analysis*. Testament to the renewed respect for Freud's ideas is the jour-nal's editorial advisory board, populated by a who's who of experts in contemporary behavioral neuroscience, including Antonio R. Dama-sio, Kandel, Joseph E. LeDoux, Benjamin Libet, Jaak Panksepp, Vilaya-nur S. Ramachandran, Daniel L. Schacter, and Wolf Singer.

Together these researchers are forging what Kandel calls a "new intellectual framework for psychiatry." Within this framework, it appears that Freud's broad brushstroke organization of the mind is destined to play a role similar to the one Darwin's theory of evolution served for molecular genetics—a template on which emerging details can be coherently arranged. At the same time, neuroscientists are uncovering proof for some of Freud's theories and are teasing out the mechanisms behind the mental processes he described.

WHEN FREUD INTRODUCED the central notion that most mental processes that determine our everyday thoughts, feelings and voli-tions occur unconsciously, his contemporaries rejected it as impossi-ble. But today's findings are confirming the existence and pivotal role of unconscious mental processing. For example, the behavior of patients who are unable to consciously remember events that occurred after damage to certain memory-encoding structures of their brains is clearly influenced by the "forgotten" events. Cogni-tive neuroscientists make sense of such cases by delineating different memory systems that process information "explicitly" (consciously) and "implicitly" (unconsciously). Freud split memory along just these lines.

Neuroscientists have also identified unconscious memory systems that mediate emotional learning. In 1996 at New York University,

LeDoux demonstrated the existence under the conscious cortex of a neuronal pathway that connects perceptual information with the primitive brain structures responsible for generating fear responses. Because this pathway bypasses the hippocampus—which generates conscious memories—current events routinely trigger unconscious remembrances of emotionally important past events, causing conscious feelings that seem irrational, such as "Men with beards make me uneasy."

Neuroscience has shown that the major brain structures essential for forming conscious (explicit) memories are not functional during the first two years of life, providing an elegant explanation of what Freud called infantile amnesia. As Freud surmised, it is not that we forget our earliest memories; we simply cannot recall them to consciousness. But this inability does not preclude them from affecting adult feelings and behavior. One would be hard-pressed to find a developmental neurobiologist who does not agree that early experiences, especially between mother and infant, influence the pattern of brain connections in ways that fundamentally shape our future personality and mental health. Yet none of these experiences can be consciously remembered. It is becoming increasingly clear that a good deal of our mental activity is unconsciously motivated.

Even if we are mostly driven by unconscious thoughts, this does not prove anything about Freud's claim that we actively repress unpalatable information. But case studies supporting that notion are beginning to accumulate. The most famous one comes from a 1994 study of "anosognosic" patients by behavioral neurologist Ramachandran of the University of California, San Diego. Damage to the right parietal region of these people's brains makes them unaware of gross physical defects, such as paralysis of a limb. After artificially activating the right hemisphere of one such patient, Ramachandran observed that she suddenly became aware that her left arm was paralyzed—and that it had been paralyzed continuously since she had suffered a stroke eight days before. This showed that she was capable of recognizing her deficits and that she had unconsciously registered these deficits for the previous eight days, despite her conscious denials during that time that there was any problem.

Significantly, after the effects of the stimulation wore off, the woman not only reverted to the belief that her arm was normal, she also forgot the part of the interview in which she had acknowledged that the arm was paralyzed, even though she remembered every other detail about the interview. Ramachandran concluded: "The remarkable theoretical implication of these observations is that memories can indeed be selectively repressed. . . . Seeing [this patient] convinced me, for the first time, of the reality of the repression phenomena that form the cornerstone of classical psychoanalytical theory."

Like "split-brain" patients, whose hemispheres become unlinked—made famous in studies by the late Nobel laureate Roger W. Sperry of the California Institute of Technology in the 1960s and 1970s—anosognosic patients typically rationalize away unwelcome facts, giving plausible but invented explanations of their unconsciously motivated actions. In this way, Ramachandran says, the left hemisphere manifestly employs Freudian "mechanisms of defense."

Analogous phenomena have now been demonstrated in people with intact brains too. As neuropsychologist Martin A. Conway of Durham University in England pointed out in a 2001 commentary in *Nature*, if significant repression effects can be generated in average people in an innocuous laboratory setting, then far greater effects are likely in real-life traumatic situations.

FREUD WENT EVEN FURTHER, though. He said that not only is much of our mental life unconscious and withheld but that the repressed part of the unconscious mind operates according to a different principle than the "reality principle" that governs the conscious ego. This type of unconscious thinking is "wishful"—and it blithely disregards the rules of logic and the arrow of time.

If Freud was right, then damage to the inhibitory structures of the brain (the seat of the "repressing" ego) should release wishful, irrational modes of mental functioning. This is precisely what has been observed in patients with damage to the frontal limbic region, which controls critical aspects of self-awareness. Subjects display a striking syndrome known as Korsakoff's psychosis: they are unaware that they are amnesic and therefore fill the gaps in their memory with fabricated stories known as confabulations.

Durham neuropsychologist Aikaterini Fotopoulou recently studied a patient of this type in my laboratory. The man failed to recall, in each fifty-minute session held in my office on twelve consecutive days, that he had ever met me before or that he had undergone an operation to remove a tumor in his frontal lobes that caused his amnesia. As far as he was concerned, there was nothing wrong with him. When asked about the scar on his head, he confabulated wholly implausible explanations: he had undergone dental surgery or a coronary bypass operation. In reality, he had indeed experienced these procedures—years before—and unlike his brain operation, they had successful outcomes.

Similarly, when asked who I was and what he was doing in my lab, he variously said that I was a colleague, a drinking partner, a client consulting him about his area of professional expertise, a teammate in a sport that he had not participated in since he was in college decades earlier, or a mechanic repairing one of his numerous sports cars (which he did not possess). His behavior was consistent with these false beliefs too: he would look around the room for his beer or out the window for his car.

What strikes the casual observer is the wishful quality of these false notions, an impression that Fotopoulou confirmed objectively through quantitative analysis of a consecutive series of 155 of his confabulations. The patient's false beliefs were not random noise—they were generated by the "pleasure principle" that Freud maintained was central to unconscious thought. The man simply recast reality as he wanted it to be. Similar observations have been reported by others, such as Martin Conway of Durham and Oliver Turnbull of the University of Wales. These investigators are cognitive neuroscientists, not psychoanalysts, yet they interpret their findings in Freudian terms, claiming in essence that damage to the frontal limbic region that produces confabulations impairs cognitive control mechanisms that underpin normal reality monitoring and releases from inhibition the implicit wishful influences on perception, memory and judgment.

Freud argued that the pleasure principle gave expression to primitive, animal drives. To his Victorian contemporaries, the implication that human behavior was at bottom governed by urges that served no higher purpose than carnal self-fulfillment was downright scandalous. The moral outrage waned during subsequent decades, but Freud's concept of man-as-animal was pretty much sidelined by cognitive scientists.

Now it has returned. Neuroscientists such as Donald W. Pfaff of the Rockefeller University and Jaak Panksepp of Bowling Green State University believe that the instinctual mechanisms that govern human motivation are even more primitive than Freud imagined. We share basic emotional-control systems with our primate relatives and with all mammals. At the deep level of mental organization that Freud called the id, the functional anatomy and chemistry of our brains is not much different from that of our favorite barnyard animals and household pets.

Modern neuroscientists do not accept Freud's classification of human instinctual life as a simple dichotomy between sexuality and aggression, however. Instead, through studies of lesions and the effects of drugs and artificial stimulation on the brain, they have identified at least four basic mammalian instinctual circuits, some of which overlap. They are the "seeking" or "reward" system (which motivates the pursuit of pleasure); the "anger-rage" system (which governs angry aggression but not predatory aggression); the "fear-anxiety" system; and the "panic" system (which includes complex instincts such as those that govern social bonding). Whether other instinctual forces exist, such as a rough-and-tumble "play" system, is also being investigated. All these brain systems are modulated by specific neurotransmitters, chemicals that carry messages between the brain's neurons.

The seeking system, regulated by the neurotransmitter dopamine, bears a remarkable resemblance to the Freudian "libido." According to Freud, the libidinal or sexual drive is a pleasure-seeking system that energizes most of our goal-directed interactions with the world. Modern research shows that its neural equivalent is heavily implicated in almost all forms of craving and addiction. It is interesting to note that Freud's early experiments with cocaine—mainly on himself—convinced him that the libido must have a specific neurochemical foundation. Unlike his successors, Freud saw no reason for antagonism between psychoanalysis and psychopharmacology. He enthusiastically anticipated the day when "id energies" would be controlled directly by "particular chemical substances." Today treatments that integrate psychotherapy with psychoactive medications are widely recognized as the best approach for many disorders. And brain imaging shows that talk therapy affects the brain in similar ways to such drugs.

Freud's ideas are also reawakening in sleep and dream science. His dream theory—that nighttime visions are partial glimpses of unconscious wishes—was discredited when rapid-eye-movement (REM) sleep and its strong correlation with dreaming were discovered in the 1950s. Freud's view appeared to lose all credibility when investigators in the 1970s showed that the dream cycle was regulated by the pervasive brain chemical acetylcholine, produced in a "mindless" part of the brain stem. REM sleep occurred automatically, every ninety minutes or so, and was driven by brain chemicals and structures that had nothing to do with emotion or motivation. This discovery implied that dreams had no meaning; they were simply stories concocted by the higher brain to try to reflect the random cortical activity caused by REM.

But more recent work has revealed that dreaming and REM sleep are dissociable states, controlled by distinct, though interactive, mechanisms. Dreaming turns out to be generated by a network of structures centered on the forebrain's instinctual-motivational circuitry. This discovery has given rise to a host of theories about the dreaming brain, many strongly reminiscent of Freud's. Most intriguing is the observation that others and I have made that dreaming stops completely when certain fibers deep in the frontal lobe have been severed—a symptom that coincides with a general reduction in motivated behavior. The lesion is exactly the same as the damage that was deliberately produced in prefrontal leukotomy, an outmoded surgical procedure that was once used to control hallucinations and delusions. This operation was replaced in the 1960s by drugs that dampen dopamine's activity in the same brain systems. The seeking system, then, might be the primary generator of dreams. This possibility has become a major focus of current research.

If the hypothesis is confirmed, then the wish-fulfillment theory of dreams could once again set the agenda for sleep research. But even if other interpretations of the new neurological data prevail, all of them demonstrate that "psychological" conceptualizations of dreaming are scientifically respectable again. Few neuroscientists still claim—as they once did with impunity—that dream content has no primary emotional mechanism.

NOT EVERYONE IS ENTHUSIASTIC about the reappearance of Freudian concepts in the mainstream of mental science. It is not easy for the older generation of psychoanalysts, for example, to accept that their junior colleagues and students now can and must subject conventional wisdom to an entirely new level of biological scrutiny. But an encouraging number of elders on both sides of the Atlantic are at least committed to keeping an open mind, as evidenced by the aforementioned eminent psychoanalysts on the advisory board of *Neuro-Psychoanalysis* and by the many graying participants in the International Neuro-Psychoanalysis Society.

For older neuroscientists, resistance to the return of psychoanalytical ideas comes from the specter of the seemingly indestructible edifice of Freudian theory in the early years of their careers. They cannot acknowledge even partial confirmation of Freud's fundamental insights; they demand a complete purge. In the words of J. Allan Hobson, a renowned sleep researcher and Harvard Medical School psychiatrist, the renewed interest in Freud is little more than unhelpful "retrofitting" of modern data into an antiquated theoretical framework. But as Panksepp said in a 2002 interview with *Newsweek* magazine, for neuroscientists who are enthusiastic about the reconciliation of neurology and psychiatry, "it is not a matter of proving Freud right or wrong, but of finishing the job."

If that job can be finished—if Kandel's "new intellectual framework for psychiatry" can be established—then the time will pass when people with emotional difficulties have to choose between the talk therapy of psychoanalysis, which may be out of touch with modern evidence-based medicine, and the drugs prescribed by psychopharmacology, which may lack regard for the relation between the brain chemistries it manipulates and the complex real-life trajectories that culminate in emotional distress. The psychiatry of tomorrow promises to provide patients with help that is grounded in a deeply integrated understanding of how the human mind operates.

Whatever undreamed-of therapies the future might bring, patients can only benefit from better knowledge of how the brain really works. As modern neuroscientists tackle once more the profound questions of human psychology that so preoccupied Freud, it is gratifying to find that we can build on the foundations he laid, instead of having to start all over again. Even as we identify the weak points in Freud's far-reaching theories, and thereby correct, revise and supplement his work, we are excited to have the privilege of finishing the job.

Dining with Robots

FROM *THE AMERICAN SCHOLAR*

When the software engineer and writer Ellen Ullman decides to make a recipe from Julia Child's Art of French Cooking, *she winds up contemplating the pleasures a robot cannot experience— and worries that even humans may be losing contact with those pleasures as well.*

On the first day of the first programming course I ever took, the instructor compared computer programming to creating a recipe. I remember he used the example of baking a cake. First you list the ingredients you'll need—flour, eggs, sugar, butter, yeast—and these, he said, are like the machine resources the program will need in order to run. Then you describe, in sequence, in clear declarative language, the steps you have to perform to turn those ingredients into a cake. Step one: Preheat the oven. Two: Sift together dry ingredients. Three: Beat the eggs. Along the way were decisions he likened to the if/then/else branches of a program: If using a counter-top electric mixer, then beat three minutes; else if using a hand electric mixer, then beat four; else beat five. And there was a reference he described as a sort of subroutine: Go to page 117 for details about varieties of yeast (with "return here" implied). He even drew a flow chart that took the recipe all the way through to the end: Let cool, slice, serve.

I remember nothing, however, about the particulars of the cake itself. Was it angel food? Chocolate? Layered? Frosted? At the time, 1979 or 1980, I had been working as a programmer for more than a year, self-taught, and had yet to cook anything more complicated than poached eggs. So I knew a great deal more about coding than about cakes. It didn't occur to me to question the usefulness of comparing something humans absolutely must do to something machines never do: that is, eat.

In fact, I didn't think seriously about the analogy for another twenty-five years, not until a blustery fall day in San Francisco, when I was confronted with a certain filet of beef. By then I had learned to cook. (It was that or a life of programmer food: pizza, takeout, whatever's stocked in the vending machines.) And the person responsible for the beef encounter was a man named Joe, of Potter Family Farms, who was selling "home-raised and butchered" meat out of a stall in the newly renovated Ferry Building food hall.

The hall, with its soaring, arched windows, is a veritable church of food. The sellers are small, local producers; everything is organic, natural, free-range; the "baby lettuces" are so young one should perhaps call them "fetal"—it's that sort of place. Before shopping, it helps to have a glass of wine, as I had, to prepare yourself for the gasping shock of the prices. Sitting at a counter overlooking the bay, watching ships and ferries ply the choppy waters, I'd sipped down a nice Pinot Grigio, which had left me with lowered sales resistance by the time I wandered over to the Potter Farms meat stall. There Joe greeted me and held out for inspection a large filet—"a beauty," he said. He was not at all moved by my remonstrations that I eat meat but rarely cook it. He stood there as a man who had—personally—fed, slaughtered, and butchered this cow, and all for me, it seemed. I took home the beef.

I don't know what to do with red meat. There is something

appalling about meat's sheer corporeality—meat meals are called *fleishidich* in Yiddish, a word that doesn't let you forget that what you are eating is *flesh*. So for help I turned to *The Art of French Cooking*, Volume I, the cookbook Julia Child wrote with Louisette Bertholle and Simone Beck. I had bought this book when I first decided I would learn to cook. But I hadn't been ready for it then. I was scared off by the drawings of steer sides lanced for sirloins, porterhouses, and T-bones. And then there was all that talk of blanching, deglazing, and making a roux. But I had stayed with it, spurred on by my childhood memories of coming across Julia on her TV cooking show, when I'd be zooming around the dial early on weekend mornings and be stopped short at the sight of this big woman taking whacks at red lumps of meat. It was the physicality of her cooking that caught me, something animal and finger-painting-gleeful in her engagement with food.

And now, as rain hatched the windows, I came upon a recipe that Julia and her coauthors introduced as follows:

SAUTÉ DE BOEUF À LA PARISIENNE

[Beef Sauté with Cream and Mushroom Sauce]

This sauté of beef is good to know about if you have to entertain important guests in a hurry. It consists of small pieces of filet sautéed quickly to a nice brown outside and a rosy center, and served in a sauce. In the variations at the end of the recipe, all the sauce ingredients may be prepared in advance. If the whole dish is cooked ahead of time, be very careful indeed in its reheating that the beef does not overcook. The cream and mushroom sauce here is a French version of beef Stroganoff, but less tricky as it uses fresh rather than sour cream, so you will not run into the problem of curdled sauce.

Serve the beef in a casserole, or on a platter surrounded with steamed rice, *risotto,* or potato balls sautéed in butter. Buttered green peas or beans could accompany it, and a good red Bordeaux wine.

And it was right then, just after reading the words "a good red Bordeaux wine," that the programming class came back to me: the instructor at the board with his flow chart, his orderly procedural steps, the if/then decision branches, the subroutines, all leading to the final "let cool, slice, serve." And I knew in that moment that my long-ago instructor, like my young self, had been laughably clueless about the whole subject of cooking food.

If you have to entertain important guests.

A nice brown outside.

Rosy center.

Stroganoff.

Curdled.

Risotto.

Potato balls in butter.

A good red Bordeaux.

I tried to imagine the program one might write for this recipe. And immediately each of these phrases exploded in my mind. How to tell a computer what "important guests" are? And how would you explain what it means to "have to" serve them dinner (never mind the yawning depths of "entertain")? A "nice brown," a "rosy center": you'd have to have a mouth and eyes to know what these mean, no matter how well you might translate them into temperatures. And what to do about "Stroganoff," which is not just a sauce but a noble family, a name that opens a chain of association that catapults the human mind across seven centuries of Russian history? I forced myself to abandon that line of thought and stay in the smaller realm of sauces made with cream, but this inadvertently opened up the entire subject of the chemistry of lactic proteins, and why milk curdles. Then I wondered how to explain "risotto": the special short-grained rice, the select regions on earth where it grows, opening up endlessly into questions of agriculture, its arrival among humans, the way it changed the earth. Next came the story of potatoes, that Inca food, the brutalities through which it arrives on a particular plate before a particular woman in Europe, before our eponymous Parisienne: how it is converted into a little round ball, and then, of course, buttered. (Then, lord help me, this brought up the whole subject of the French and butter, and how can they possibly get away with eating so much of it?)

But all of this was nothing compared to the cataclysm created by "a good red Bordeaux."

The program of this recipe expanded infinitely. Subroutine opened from subroutine, association led to exploding association. It seemed absurd even to think of describing all this to a machine. The filet, a beauty, was waiting for me.

Right around the time my programming teacher was comparing coding to cake-making, computer scientists were finding themselves stymied in their quest to create intelligent machines. Almost from the moment computers came into existence, researchers believed that the machines could be made to think. And for the next thirty or so years, their work proceeded with great hope and enthusiasm. In 1967 the influential MIT computer scientist Marvin Minsky declared, "Within a generation the problem of creating 'artificial intelligence' will be substantially solved." But by 1982 he was less sanguine about the prospects, saying, "The AI problem is one of the hardest science has ever undertaken."

Computer scientists had been trying to teach the computer what human beings know about themselves and the world. They wanted to create inside the machine a sort of mirror of our existence, but in a form a computer could manipulate: abstract, symbolic, organized according to one theory or another of how human knowledge is structured in the brain. Variously called "micro worlds," "problem spaces," "knowledge representations," "classes," and "frames," these abstract universes contained systematized arrangements of facts, along with rules for operating upon those—theoretically all that a machine would need to become intelligent. Although it wasn't characterized as such at the time, this quest for a symbolic representation of reality was oddly Platonic in motive, a computer scientist's idea of the pure, unchanging forms that lie behind the jumble of the physical world.

But researchers eventually found themselves in a position like mine when trying to imagine the computer program for my *boeuf à la Parisienne:* the network of associations between one thing and the next simply exploded. The world, the actual world we inhabit, showed itself to be too marvelously varied, too ragged, too linked and inter-connected, to be sorted into any set of frames or classes or problem spaces. What we hold in our minds is not abstract, it turned out, not an ideal reflection of existence, but something inseparable from our embodied experience of moving about in a complicated world.

Hubert L. Dreyfus, a philosopher and early critic of artificial intel-ligence research, explained the problem with the example of a simple object like a chair. He pointed out the futility of trying to create a symbolic representation of a chair to a computer, which had neither a body to sit in it nor a social context in which to use it. "Chairs would not be equipment for sitting if our knees bent backwards like those of flamingoes, or if we had no tables, as in traditional Japan or the Aus-tralian bush," he wrote in his 1979 book *What Computers Can't Do.* Letting flow the myriad associations that radiate from the word *chair,* Dreyfus went on:

> Anyone in our culture understands such things as how to sit on kitchen chairs, swivel chairs, folding chairs; and in arm chairs, rock-ing chairs, deck chairs, barber's chairs, sedan chairs, dentist's chairs, basket chairs, reclining chairs . . . since there seems to be an indefi-nitely large variety of chairs and of successful (graceful, comfort-able, secure, poised, etc.) ways to sit in them. Moreover, understanding chairs also includes social skills such as being able to sit appropriately (sedately, demurely, naturally, casually, sloppily, provocatively, etc.) at dinners, interviews, desk jobs, lectures, audi-tions, concerts. . . .

At dinners where one has to entertain important guests . . . at the last minute . . . serving them beef in a French version of Stroganoff . . . with buttered potatoes . . . and a good red Bordeaux.

SEVERAL WEEKS after making Julia's *boeuf,* I was assembling twelve chairs (dining chairs, folding chairs, desk chair) around the dining table, and I was thinking not of Dreyfus but of my mother. In her younger days, my mother had given lavish dinner parties, and it was she who had insisted, indeed commanded, that I have all the necessary equipment for the sort of sit-down dinner I was giving that night. I surveyed the fancy wedding-gift stainless she had persuaded me to register for ("or else you'll get a lot of junk," she said), the Riedel wine glasses, also gifts, and finally the set of china she had given me after my father's death, when she sold their small summer house—"the country dishes" is how I think of them, each one hand-painted in a simple design, blue cornflowers on white.

It wasn't until I started setting the table, beginning with the forks, that I thought of Dreyfus. Salad forks, fish forks, crab forks, entrée forks, dessert forks—at that moment it occurred to me that the paradigm for an intelligent machine had changed, but what remained was the knotty problem of teaching a computer what it needed to know to achieve sentience. In the years since Dreyfus wrote his book, computer scientists had given up on the idea of intelligence as a purely abstract proposition—a knowledge base and a set of rules to operate upon it—and were now building what are called social robots, machines with faces and facial expressions, who are designed to learn about the world the way human beings do: by interacting with other human beings. Instead of being born with a universe already inscribed inside them, these social machines will start life with only basic knowledge and skills. Armed with cute faces and adorable expressions, like babies, they must inspire humans to teach them about the world. And in the spirit of Dreyfus, I asked myself: If such a robot were coming to dinner, how could I, as a good human hostess and teacher, explain everything I would be placing before it tonight?

Besides the multiple forks, there will be an armory of knives: salad knife, fish knife, bread knife, dessert knife. We'll have soup spoons and little caviar spoons made of bone, tea spoons, tiny demitasse spoons, and finally the shovel-shaped ice-cream spoons you can get only in Germany—why it is that only Germans recognize the need for this special ice cream implement? My robot guest could learn in an instant the name and shape and purpose of every piece of silverware,

I thought; it would instantly understand the need for bone with caviar because metal reacts chemically with roe. But its mouth isn't functional; the mouth-part is there only to make us humans feel more at ease; my robot guest doesn't eat. So how will it understand the complicated interplay of implement, food, and mouth—how each tool is designed to hold, present, complement the intended fish or vegetable, liquid or grain? And the way each forkful or spoonful finds its perfectly dimensioned way into the moist readiness of the mouth, where the experience evanesces (one hopes) into the delight of taste?

And then there will be the wine glasses: the flutes for champagne, the shorter ones for white wine, the pregnant Burgundy glasses, the large ones for Cabernet blends. How could I tell a machine about the reasons for these different glasses, the way they cup the wine, shape the smell, and deliver it to the human nose? And how to explain wine at all? You could spend the rest of your life tasting wine and still not exhaust its variations, each bottle a little ecosystem of grapes and soils and weather, yeast and bacteria, barrels of wood from trees with their own soil and weather, the variables cross-multiplying until each glassful approaches a singularity, a moment in time on earth. Can a creature that does not drink or taste understand this pleasure? A good red Bordeaux!

I went to the hutch to get out the china. I had to move aside some of the pieces I never use: the pedestaled cigarette holders, the little ashtrays, the relish tray for the carrots, celery, and olives it was once de rigueur to put on the table. Then I came to the coffeepot, whose original purpose was not to brew coffee—that would have been done in a percolator—but to serve it. I remembered my mother presiding over the many dinners she had given, the moment when the table was scraped clean of crumbs and set for dessert, the coffee cups and saucers stacked beside her as she poured out each cup and passed it down the line. Women used to serve coffee at table, I thought. But my own guests would walk over and retrieve theirs from the automatic drip pot. My mother is now ninety-one; between her time as a hostess and mine, an enormous change had occurred in the lives of women. And, just then, it seemed to me that all that upheaval was contained in the silly fact of how one served coffee to dinner guests. I knew I would never want to go back to Mother's time, but all the same I suddenly missed the world of her dinner parties, the guests waving their cigarettes as they chatted, my mother so dressed up, queenly by the coffee pot, her service a kind of benign rule over the table. I put the pot in the corner of the hutch and thought: It's no good trying to explain all this to my robot guest. The chain of associations from just this one piece of china has led me to regret and nostalgia, feelings I can't explain even to myself.

THE REAL PROBLEM with having a robot to dinner is pleasure. What would please my digital guest? Human beings need food to survive, but what drives us to choose one food over another is what I think of as the deliciousness factor. Evolution, that good mother, has seen fit to guide us to the apple instead of the poison berry by our attraction to the happy sweetness of the apple, its fresh crispness, and, in just the right balance, enough tartness to make it complicated in the mouth. There are good and rational reasons why natural selection has made us into creatures with fine taste discernment—we can learn what's good for us and what's not. But this very sensible survival imperative, like the need to have sex to reproduce, works itself out through the not very sensible, wilder part of our nature: desire for pleasure.

Can a robot desire? Can it have pleasure? When trying to decide if we should confer sentience upon another creature, we usually cite the question first posed by the philosopher Jeremy Bentham: Can it suffer? We are willing to ascribe a kind of consciousness to a being whose suffering we can intuit. But now I wanted to look at the opposite end of what drives us, not at pain but at rapture: Can it feel pleasure? Will we be able to look into the face of a robot and understand that some deep, inherent need has driven it to seek a particular delight?

According to Cynthia Breazeal, who teaches at MIT and is perhaps the best known of the new social-robot researchers, future digital creatures will have drives that are analogous to human desires but that will have nothing to do with the biological imperatives of food and sex. Robots will want the sort of things that machines need: to stay in good running order, to maintain physical homeostasis, to get the attention of human beings, upon whom they must rely, at least until they learn to take care of themselves. They will be intelligent and happy the way dolphins are: in their own form, in their own way.

Breazeal is very smart and articulate, and her defense of the eventual beingness of robotic creatures is a deep challenge to the human idea of sentience. She insists that robots will eventually become so lifelike that we will one day have to face the question of their inherent rights and dignity. "We have personhood because it's granted to us by society," she told me. "It's a status granted to one another. It's not innately tied to being a carbon-based life form."

So challenged, I spent a long time thinking about the interior life of a robot. I tried to imagine it: the delicious swallowing of electric current, the connoisseurship of voltages, exquisite sensibilities sensing tiny spikes on the line, the pleasure of a clean, steady flow. Perhaps the current might taste of wires and transistors, capacitors and rheostats, some components better than others, the way soil and water make up the *terroir* of wine, the difference between a good Bordeaux and a middling one. I think robots will delight in discerning patterns, finding mathematical regularities, seeing a world that is not mysterious but beautifully self-organized. What pleasure they will take in being fast and efficient—to run without cease!—humming along by their picosecond clocks, their algorithms compact, elegant, error-free. They will want the interfaces between one part of themselves and another to be defined, standardized, and modular, so that an old part can be unplugged, upgraded, and plugged back in their bodies forever renewed. Fast, efficient, untiring, correct, standardized, organized: the virtues we humans strive for but forever fail to achieve, the reasons we invented our helpmate, the machine.

THE DINNER PARTY, which of course proceeded without a single robot guest, turned out to be a fine, raucous affair, everyone talking and laughing, eating and drinking to just the right degree of excess.

And when each guest rose to pour his or her own cup of coffee, I knew it was one of those nights that had to be topped off with a good brandy. By the time the last friend had left, it was nearly two a.m., the tablecloth was covered with stains, dirty dishes were everywhere, the empty crab shells were beginning to stink, and the kitchen was a mess. Perfect.

Two days later I was wheeling a cart through the aisles at Safeway—food shopping can't always be about fetal lettuces—and I was thinking how neat and regular the food looked. All the packaged, pre-prepared dinners lined up in boxes on the shelves. The meat in plastic-wrapped trays, in standard cuts, arranged in orderly rows. Even the vegetables looked cloned, identical bunches of spinach and broccoli, perfectly green, without an apparent speck of dirt. Despite the influence of Julia Child and California-cuisine guru Alice Waters, despite the movement toward organic, local produce, here it all still was: manufactured, efficient, standardized food.

But of course it was still here, I thought. Not everyone can afford the precious offerings of the food hall. And even if you could, who really has the time to stroll through the market and cook a meal based on what looks fresh that day? I have friends who would love to spend rainy afternoons turning a nice filet into *boeuf à la Parisienne*. But even they find their schedules too pressed these days; it's easier just to pick something up, grab a sauce out of a jar. Working long hours, our work life invading home life through e-mail and mobile phones, we all need our food-gathering trips to be brief and organized, our time in the kitchen efficiently spent, our meals downed in a hurry.

As I picked out six limes, not a bruise or blemish on them, it occurred to me that I was not really worried about robots becoming sentient, human, indistinguishable from us. That long-standing fear—robots who fool us into taking them for humans—suddenly seemed a comic-book peril, born of another age, as obsolete as a twenty-five-year-old computer.

What scared me now were the perfect limes, the five varieties of apples that seemed to have disappeared from the shelves, the dinner

I'd make and eat that night in thirty minutes, the increasing rarity of those feasts that turn the dining room into a wreck of sated desire. The lines at the check-out stands were long; neat packages rode along on the conveyor belts; the air was filled with the beep of scanners as the food, labeled and bar-coded, identified itself to the machines. Life is pressuring us to live by the robots' pleasures, I thought. Our appetites have given way to theirs. Robots aren't becoming us, I feared; we are becoming them.

—in memory of Julia Child

Andrea Barrett

The Sea of Information

FROM *THE KENYON REVIEW*

> *Reflecting on the mysterious interplay of fact and imagination, the novelist Andrea Barrett illuminates the differences in the way that artists and scientists work, especially in their use of information.*

I've always thought of myself as a writer more than usually dependent on news from the outside world. My imagination is nourished by old books, old bones, fossils, feathers, paintings, photographs, museums of every kind and size, microscopes and telescopes, plants and birds; I like to learn things and I thought—I still do think, although my ideas have darkened—that all this information feeds my fiction.

It wasn't so strange, then, to find myself excited by the slim gray book stamped "Property of the City of New York" and titled *What You Should Know about TUBERCULOSIS*, which fell into my hands sometime during 1999. Inside it I found photographs of children deformed by tuberculosis of the spine, of a young man perched on a tenement roof, gazing at the tattered tent and cot in which he is "taking the cure at home, in summer," of a young woman on a similar roof, bundled in mittens and a thick coat, smiling as she sweeps the snow from around her tent while curing "on the roof in winter." Copies of this book, I learned, had been passed out to public high school students in New York City between 1910 and 1920, a place and a period I knew almost nothing about. Still, I found myself wondering what it might have been like to be one of the students, studying the book and preparing to answer the test questions at the end. Pausing, maybe, over a clumsy drawing of a tree being attacked by an ax; noticing that the trunk was labeled "Tuberculosis" and the many branches bore such names as "incapacitated workers," "thousands dependent on charity," and "hunchbacks," while the tangled roots included "poverty," "child labor," and "careless consumptives."

Perhaps that student might have flipped between the Tuberculosis Tree and the map of a nearby neighborhood, the blocks bounded by the Bowery, East Houston, Canal, East Broadway, and Columbia. Within that area, every building sheltering a known case of tuberculosis was marked by a black dot. There were a great number of dots; some blocks were almost entirely black and anyone living there would have known tuberculosis intimately. Everywhere people suffered while visitors, often unwelcome, descended upon them. Doctors, charity workers, public health officials. Visiting nurses, trained by a 1915 handbook to think that:

In the first place, tuberculosis is largely a disease of the poor—of those on or below the poverty line. We must further realize that there are two sorts of poor people—not only those financially handicapped and so unable to control their environment, but those who are mentally and morally poor, and lack intelligence, will power, and self-control. The poor, from whatever cause, form a class whose environment is difficult to alter. And we must further realize that these patients are surrounded in their homes by people of their own kind—their families and friends—who are also poor. It is this fact which makes the task so difficult, and makes the prevention and cure of a preventable and curable disease a matter of utmost complexity.[1]

The *sound* of that language—the officious, pushy, condescending *sound* of that—along with the eerie photographs and the remarkable drawing of the Tuberculosis Tree, made me want to write a novel. The feeling was as sudden, as intense, and as irrational as falling in love.

Who can say why we're drawn to one person, or one subject? All I knew was that I was, and that the impulse was powerful. The novel would be set, I thought, in a public sanatorium; I imagined a sort of working-class, American version of Thomas Mann's *The Magic Mountain*, which is set at a fancy private sanatorium in the Alps in the years just before World War I erupted throughout Europe. Mine would start a little later—but before the American entry into the war, I thought; I didn't want to have to deal with the war, which seemed like a whole other subject—and would explore not the lives of the rich and idle, but those of the working poor. A friend to whom I tried to describe it jokingly called it "The Magic Molehill."

Several things happened to that impulse, though, along the way.

One is that I used a description of that little gray book, and the world from which it had come, as part of a proposal, by means of which I hoped to gain a fellowship at the New York Public Library. There I planned to research the novel, and to begin writing it in congenial surroundings. A library seemed like the perfect temporary home for me. Over the past decade or so I've written about subjects as diverse as China during the Cultural Revolution, the evolutionary biologist Alfred Russel Wallace in the Malay Archipelago, the monk and botanist Gregor Mendel and his experiments with peas, nineteenth-century Arctic exploration, surveying and mapmaking in the 1860s, and the development of paleontology in nineteenth-century America. I've studied early conceptions of the formation of dew and rain, the founding of the utopian colony of New Harmony, and the manufacture of the glass eyes used by taxidermists to simulate life in stuffed game.

Ridiculous, I know. But by now, despite that, I've learned to follow where the spark of interest leads me. Often I'm drawn toward the past, and toward material involving natural history and the sciences; I can't explain that either, except to say that for me this material seems full of potential, charged and fresh and inspiring. Sometimes the stories *lead* me to the research: halfway into a sentence I may realize I don't have any idea whether a character climbing in the Karakorum range in the 1850s would be wearing hobnailed boots or metal crampons, and then it's time to study the history of mountaineering. Sometimes, as with the little gray book, I bump into things by accident, and they in turn either lead to stories, or bend stories already in progress into a different shape. But always, libraries have been my most dependable resource for this material. At the enormous library in New York, I imagined I'd find all sorts of secret treasures, as indeed I did. But I also found myself strangely lost, and in unknown territory.

BECAUSE I WAS GRANTED THAT FELLOWSHIP, my husband and I moved to New York City for a year. During that July and August I walked around the Lower East Side, looking at tenement buildings marked on the map in the gray book. In the Williamsburg section of Brooklyn, where we were living, I walked around the waterfront and inspected the shattered docks from which a largely immigrant population had stepped onto ferries headed for Manhattan. I walked around an old sugar refinery, hardly changed from the early twentieth century and still belching steam, in which I decided I'd have one of my characters work. I read Walt Whitman's "Crossing Brooklyn Ferry," and I flipped through old maps of the neighborhood, and I looked forward to the start of the fellowship itself, to the nine months during which I'd occupy a little office, grouped with fourteen other offices around a common area where I and my fellow Fellows might gather to share ideas.

We were a mixed group, I knew: four fiction writers, one poet, the remainder historians, biographers, and critics, studying everything from Ben Franklin's attitudes toward slavery to Russian iconography to the social history of an extended African-American family. I knew that the scholars were also excited by books and maps and unusual facts, although I had little sense, then, of the different ways in which we'd actually use the materials we found. We'd have access, I knew, to virtually any source we could imagine; fill out a call slip, and it would appear. This part I imagined as heavenly, and that turned out to be true. But what also turned out to be true was that the first day of our fellowship was September 10, 2001. We had our pictures taken, we met the staff, we got ID cards and had a tour. The next day, we were told, would be our first real introduction to the library's resources.

That morning I woke early, gathered my papers together, dressed, and then walked our dog to the dog-sitters' place in an old garage beside the East River, directly across from Lower Manhattan. On my way back to pick up my briefcase and head for the subway, I heard a thump and turned to see smoke and flames pouring out of one of the towers that had dominated our view since we'd moved there. Later, from the roof of our building, I saw the towers burn and fall.

This isn't meant to be an exploration of the effects of September 11, and so I won't include any more details about the next few weeks except to say that, unlike so many others, my husband and I were extraordinarily fortunate. Neither of us was hurt, nor were family members, nor close friends. We were bystanders, nothing more. But even for us, as for everyone, everything was different after that day.

Like many writers, I found myself unable to write in the aftermath of the attacks—not just fiction, but anything. What was the point? I thought at first, and for a long time after. But of course there's always a point: reading and writing are two of the ways we make sense of our mysterious, sometimes terrible, world. There were reasons why, all through the autumn months, the Main Reading Room at the Library was packed with people reading newspapers and books, searching for material on-line, talking to each other and to the librarians. Eventually I took my cue from those people pouring in off the street. It was through reading, which grew into a more directed kind of research, that I first began to try to grapple with what had happened.

I wrote, haltingly, an autobiographical essay about witnessing the attacks as a new New Yorker, but although there was some relief in trying to get a sketch of those weeks down on paper, I couldn't get at the heart of things; the experience was still too raw, and some of it was still going on. Anthrax was floating through the mails; subway trains were mysteriously stopped, rerouted, evacuated; buildings where I went to meet friends would be suddenly sealed, surrounded by armed men and flashing lights. The newspapers were filled with rumors and fresh horrors, with heart-wrenching photographs and rhetorical excess and calls for war. I found myself driven not toward these renderings of what surrounded me but away, toward other times and places in which similar events—events that felt analogous—had taken place. More and more I found myself delving not into what I'd once thought was the central subject matter of my novel, but into what had been going on around my characters in their sanatorium, which was World War I.

If novelists think, perhaps this is *how* we think: through a frenzy of metaphor-making and analogy-building, an accretion of meaningful images juxtaposed in ways that seem to us fruitful, although to someone else they might seem baffling. On different days (or sometimes on the same day, when I was feeling particularly lost), I read about the history of Brooklyn in the early twentieth century; the process of sugar refining; immigration policy; the histories of nursing, dispensaries, and public health; what it was like to live in Russian Poland at the end of the nineteenth century and the beginning of the twentieth; the Russian Revolution; the experience of Americans who enlisted with Canadian or British regiments during the early years of the Great War; the American response to the war before entering it; the military draft and how it was implemented; the uses of such chemical and biological weapons as gas and anthrax, both of which were used for the first time then; and the polio epidemic that started in Brooklyn during the summer of 1916.

I read, with a kind of terrible fascination, about a night in the middle of that same summer, when what had once been a munitions dump on Black Tom Island, warehouses and boxcars filled with shells and powder made here and waiting to be shipped to France, blew up in an explosion so massive that windows blew out across Lower Manhattan and New Jersey, the ground shook in Brooklyn, and the island itself disappeared. German spies were blamed for that; German-Americans were blamed for having sheltered the spies. The waves of anti-German prejudice that followed helped shift American sentiment from neutrality toward participation in the war.

Why was I reading all this? Why do all this work, especially when I wasn't writing and didn't know if, when I started again, I'd find a way to use any of it? And especially when I might more usefully have been out in the world, helping someone, fixing something: cleaning up the rubble or raising money or aiding the families of the dead. Instead I read, which is what I do. I read like that—I have always read like that—because it's the only way I know to deeply inhabit a world other than the limited one of my own experience. It's the way I sink into the hearts and minds of invented characters, who incarnate themselves in the odd intersections of apparently disparate fields, and who then, if I'm lucky, manage to understand and articulate what I cannot. Reading, which gives me access to lives I haven't lived, am not living, probably won't live, is how I find my way to writing: in this case how I found my way *back* to writing.

And yet—for the first time in my life, I was surrounded by actual scholars as I read; who swam through a sea of information and marshaled facts in ways that were unfamiliar to me. As I listened to them talk about their own researches into history and culture and the intersections between what they were studying and what was going on around us that fall and winter, I was forced to think about how very differently scholars on the one hand, and poets and novelists on the other, approach their material. We not only do research differently, we do it in a different spirit, for a different purpose; and then we turn the results to different ends. My colleagues spoke about discovering new sources, about rummaging around in original sources, sometimes in several languages; about finding evidence and, from that evidence, constructing and testing hypotheses and then building chains of argument. Their process reminded me of science, seemed almost like a *kind* of science.

Because I'm so drawn to scientific material, I'm sometimes also drawn to its methods. But although I found my colleagues' approach both admirable and fascinating, I got lost when I attempted to emulate it. I've never been able to write unless I take a different, more wandering path. While brilliant fiction has been written by people who work rationally, write outlines, plan beforehand where they are going, and are able to think their way through the structure of a novel or a story, I've always had to feel my way more blindly and intuitively. There are excellent novels that *do* make arguments, and are as essentially polemical as a work of nonfiction: Dickens's *Hard Times* comes to mind. But these aren't the novels, for whatever reasons, that I most love; nor are they what I attempt to write. Usually what I'm trying to build isn't an argument, isn't overtly didactic, doesn't state its premises clearly, often doesn't operate linearly, and can't be reduced to a clear statement.

It took being in the company of this group of scholars to teach me that when I do what, up until then, I had always *thought* of as "research," I was only skimming the surface. I know a little about a lot of things, but the only thing I really know well is storytelling, in all its forms—and that's the end my so-called research serves. Although most of my fiction is set in the past and employs the materials of history, I'm not trying to discover new facts, or to develop new hypotheses, or in some sense prove "how it was." I'm trying to shape a narrative that allows a reader to *feel* what it was like to be a particular person, or set of persons, caught in a particular situation at a particular place and time. When I looked at the sugar factory in Williamsburg and dug up early photographs of it or articles about how sugar was processed there, I was looking for those details that would allow me to imagine *one person* working in and moving through that space; I was imagining it as background. A social historian, looking at the same material, might ask what wages were, what the ethnicity of the workers was, where those workers lived, who got paid what for what hours. Facts, from which inferences could be drawn. I was looking for something else.

I wanted what would help me not to tell, but to *show*. If I could convey what I wanted to convey by a set of logically ordered and clear statements, I'd do that. But a good novel or story or poem tries to convey a different kind of knowledge, and to operate on the reader in a different way, through the emotions and the senses. Facts can help *evoke* emotion, especially those that transmit texture, tonality, and sensual detail. But facts can't drive a piece. Research, no matter how compelling, may give me the bones of a fiction but never the breath and the blood. It's a wonderful, sometimes immensely useful tool that helps give me something to write about. But without the transforming force of the imagination, the result is only information.

In 1936, when a different war was looming on the horizon, Walter Benjamin wrote this:

> Every morning brings us the news of the globe, and yet we are poor in noteworthy stories. This is because no event any longer comes to us without already being shot through with explanation. In other words, by now almost nothing that happens benefits storytelling; almost everything benefits information. Actually, it is half the art of storytelling to keep a story free from explanation as one reproduces it. [...] The value of information does not survive the moment in which it was new. It lives only at that moment; it has to surrender to it completely and explain itself to it without losing any time.[2]

Information, information. I was drowning in it, one stream pouring in from the daily news of the world, the other bubbling up from the library stacks, while all around me people used it to build explanations for the present and the past. Every day I'd learn something more about the world, the war, tuberculosis, public health, propaganda. And every day I felt farther away from the writing itself. It was as if, to hark back again to *Hard Times,* I had turned while my attention had drifted into the student in Professor Gradgrind's class who, when asked to define a horse, responded:

> "Quadruped. Graminivorous. Forty teeth, namely twenty-four grinders, four eye-teeth, and twelve incisive. Sheds coat in the spring;

in marshy countries, sheds hoofs, too. Hoofs hard, but requiring to be shod with iron. Age known by marks in the mouth."[3]

Caught up in learning about the equivalent of grinders and eye-teeth, I'd forgotten that while facts may be *in* a text—sometimes delectably—they can't *be* the text itself. Slowly I began to relearn something I'd once grasped, but had lost sight of: that emotion—that central element of fiction—derives not from information or explanation, nor from a logical arrangement of facts, but specifically from powerful images and from the qualities of language: diction, rhythm, form, structure, association, metaphor. And sometimes I also had glimmers of another thing I'd once known: how effectively information can be used to wall off emotion. How the gathering of information can take the place of actual understanding. I had built, as I am only now realizing, quite a substantial wall. As if any wall could block out those two towers.

As I went longer and longer without writing fiction, the novel I wanted to write—at that point still purely hypothetical—began to seem as if it should encompass not only the sanatorium experience, but also the experience of a country on the verge of entering a war. When I did finally start writing in the autumn of 2002, around the anniversary of September 11, what I began was no longer the novel I'd first imagined. Still, I thought this had to do with a change in the *kind* of information I needed—less about tuberculosis, more about the war. More recently I've come to see that the change is of a different sort.

There's no single, central character anymore, no one prism through which everything is refracted. Instead there are a tangle of characters talking to and through and around each other, struggling to make sense both of what's happened to them personally, and what's going on in the larger world: not just the distant war but the burgeoning implements and technologies of war, the changes in politics and society. Does that sound a bit like a group of people gathered in a library during and after a crisis, talking their way through the events of the day? Perhaps it does. A war, like an epidemic, happens to

276 | ANDREA BARRETT

everyone; as autobiography, no matter how vigorously squelched, has a way of pulsing beneath the surface of one's work.

In this version of the novel, a character named Leo Marburg is among a handful of patients who, in a state sanatorium in the Adirondacks, amid a swirl of conversations about the war abroad and the contributions made to it here—chemical weapons, X-ray machines, munitions, eager volunteers—causes something to happen. It's pointless to say more about Leo or how I think the plot might unfold. Partly that's because I don't *know* yet—I can guess, and I can plan, but the experience of writing previous books suggests that I'll be wrong, and that I'll be surprised at every turn. Partly it's because plot summaries are boring. If the novel's working, I shouldn't be able to reduce it to an outline, and I shouldn't be able to articulate what it's really about. All I can say is that, partly through the experience of living in New York during that difficult year, the world of these characters keeps enlarging and the series of intersecting circles keeps widening.

The more I learned about World War I, the more I saw how much it had in common with what was known at the time as the "War against Tuberculosis": the material contained in the little gray book. Those wars overlap exactly in time—but also, more important, in their uses of propaganda and corrupted public language. The militaristic, and yet at the same time euphemistic, language of the "War against Tuberculosis" is very like that found in the documents used to whip up American support for entry into the war. The *sound* of that language interests me a good deal—it's a sound that's becoming familiar again. And it was sound—the rhythm and tone of a particular narrative voice, its diction and pacing and music—that helped me begin the actual writing. What I had experienced changed the novel; the new information I gathered was necessary to it; neither experience nor information were sufficient. What I needed was a resonant metaphoric framework, and a voice.

In the handful of pages I wrote before I had any real sense of where the novel was headed there's hardly a single complete sentence, never mind a coherent paragraph. But even from those I could get a sense of structure, intent, and a kind of verbal patterning, despite the fact that most of the nouns—the facts—were still missing. Actual phrases and sentences are mingled with lumpy directions to myself, enclosed within parentheses—and these, not a chain of reasoning, led me on. This is how the first gesture sounded on the day I sketched it out:

"That summer, everything seemed to be crumbling *(disintegrating, catching fire, happening. . . .)* all at once. *(see the NYC newspapers. Focus down a window of time—roughly the last week or two weeks of June 1916. Impending war with Mexico, news of the war overseas, all the local accidents and labor disputes, incidents of German sabotage).* A ship blew up, a train was (. . . .); in France *(insert some event from the war here),* while a (. . . .) was (. . . .) in New York Harbor. A *(. . . insert two other trivial events here, from Brooklyn newspaper).*

"In Brooklyn (. x . .) children had been paralyzed by polio on June (. y . .), and another (. x . .) on June (. z . .). By the end of the month, *(. . . summarize the panic of the epidemic here; the children being turned back by police on Long Island, the families being rousted from their apartments, etc.)* In Greenpoint and Williamsburgh, *(. . . more examples here. Then end this opening beat, which has narrowed from Europe and all over the States, to New York Harbor, to Brooklyn, and down to Leo's neighborhood, with this):*

"At the sugar refinery where Leo Marburg worked, three men had been dismissed on suspicion of sabotage after a fire broke out in the warehouse. Leo, now working more hours than usual, was exhausted. Two of his landlady's children were sick; his own head ached and the nagging cough he'd had since winter kept him up even when the children weren't crying and the ambulances weren't roaming the streets with their sirens shrieking. Each day he went in to work and *(summarize his routine duties here).*

"One Wednesday, early in July . . ."

ESSENTIALLY THAT'S AN INFORMATION-FREE PASSAGE; truly, writing can be ridiculous. And yet despite the obvious problems and omissions there's something—a kind of feeling, a structure, a tone—gesturing there. That something springs not from experience or information but from their synthesis and growth in my imagination.

Each time I try to do this, I relearn the lesson that I can't, during the process of writing, relegate imagination to an inferior place. I can't let research, my ally and comfort for so long, push its way to the head of the line. The work never comes alive until I give up the idea that I know what I'm writing about, and allow myself to be led—by the life that goes on outside us, in the world, and also by the fertile life going on in secret, inside our heads—into new and strange territory. Any text, I learn each time, is a tissue of the imagination, in which facts, if we choose to embed them, rest safely encysted.

By now, of course, that first, exploratory passage has disappeared, replaced by something that sounds almost shockingly different. But that too is part of the journey of the imagination, which the dictionary defines as: "the act or power of forming a mental image of something not present to the senses or never before wholly perceived in reality." A useful reminder that the imagination is founded in, flourishes on, *images:* pictures, fortified by sight, touch, taste, sound, and passionate emotion. One image leads me to the next, and the next; the next requires the revision of all that has come before, and on it goes.

So what is it, then, that I'm trying to say? I went to New York with an idea for a book, which was inspired by another book; the world changed while I was there, and so did I; the book I meant to write changed as well. Do I mean to say that writers should look within, or look without? That they should write from experience, or from research, or from imagination?

Yes, I would say. Not either/or, but all those things. Writing is mysterious, and it's supposed to be. Craft guides a writer at every step, as does knowledge of earlier work; we accomplish little without those foundations. Research can help, if it feeds the imagination and generates ideas; a plan is also a wonderful thing, if a writer's imagination works that way. Groping blindly, following glimmers of structure and sound, is far from the only way; other writers work differently to good effect and any path that gets you there is a good path in the end. But one true thing among all these paths is the need to tap a deep vein of connection between our own uncontrollable interior preoccupations, and what we're most concerned about in the world around us. We write in response to that world; we write in response to what we read and learn; and in the end we write out of our deepest selves, the live, breathing, bleeding place where the pictures form, and where it all begins.

1. Ellen N. LaMotte, *The Tuberculosis Nurse: Her Functions and Qualifications* (New York: G.P. Putnam's Sons, 1915). Reprinted in *From Consumption to Tuberculosis: A Documentary History,* ed. by Barbara Rosenkrantz (New York: Garland Publishers, 1994), p. 442.

2. Walter Benjamin, "The Storyteller." In *Illuminations* (New York: Schocken Books, 1969), pp. 89-90.

3. Charles Dickens, *Hard Times* (1854) (New York: Holt, Rinehart and Winston, 1961), p. 4.

DIANE ACKERMAN

Even Bees Must Rest Their Heads

FROM THE *NEW YORK TIMES*

Observing a sleeping bumblebee on a cold morning, Diane Ackerman experiences a lovely creative moment with nature.

Atop a fothergilla leaf, a fat bumblebee is slumbering, motionless. A breeze jolts the leaf, but the bee seems to be buttoned in place.

Sometimes foraging bees can't make it back to the hive before dark, and must grab a leaf and wait through night chill for dawn. Many die in the process. On cold nights, they need hive mates to keep warm, and last night was one of those.

Unlike honey bees (and humans, for that matter), bumblebees don't store food and huddle indoors, staying sociable and warm all winter. Instead, most of their family dies, leaving behind only young pregnant females that hibernate below ground, held in suspense until spring.

This is an older female (younger ones stay home to help in the nursery). Is it waiting for the first sunlight? The morning damp to dry?

At last, it tilts its head and stirs. Two feet jut out and back, in a hokey-pokey sort of move, then its wings flex a little without lifting.

Ten minutes later, still not flying, it climbs to the top of the leaf and ambles around. Then it begins working the bellows of its abdomen, rhythmically, in a bee version of calisthenics, to limber up the flight muscles. Still no flashing wings, but lots of quiver.

Bumblebees can twitch their wing muscles fast to create body heat. A neat trick. Monarch butterflies also shiver to stay warm, or use their wings as solar panels, unless the temperature really plunges, and then they're as paralyzed as this bumblebee was, and easy prey.

I remember one chilly morning in California, when a colleague and I held just-tagged monarch butterflies in our open mouths and warmed them with our breath, so that they could fly to safety.

This bumblebee's two leg baskets are bulging with pollen, a heavy load to haul home. It may have misjudged the pollen weight or the flying time.

There is much for even a bumblebee to analyze and decide.

Where is her ground hive, anyway? It's bound to be in a burrow dug by some obliging creature, or maybe a grassy nest stolen from a field mouse.

I'd go looking, but I want to keep the bumblebee company on the ledge of its day. As sun finds the fothergilla, the bumblebee flurries its wings, levitates, and lumbers away at last, buzzing with rich cargo.

About the Contributors

Poet, essayist, naturalist, DIANE ACKERMAN is the author of twenty works of poetry and nonfiction, including, most recently, *An Alchemy of Mind: The Marvel and Mystery of the Brain; Origami Bridges: Poems of Psychoanalysis and Fire;* and *Animal Sense* (children's), illustrated by Peter Sis. Dr. Ackerman has received many prizes and awards, including a Guggenheim Fellowship, the John Burroughs Nature Award, and the Lavan Poetry Prize. She has the somewhat unusual distinction of having a molecule named after her—dianeackerone (a crocodilian sex pheromone). She has also hosted a five-hour PBS television series inspired by her bestselling book, *A Natural History of the Senses.*

"I love discovering some of the ordinary miracles that surround us each day," she explains, "especially at dawn, a luminous time before all the weight of world has chance to settle on one's shoulders."

JENNIFER ACKERMAN is a writer specializing in the biological sciences. Her work aims to explore the riddle of humanity's place in the natural world, blending scientific knowledge with imaginative vision. Her books, include *Chance in the House of Fate: A Natural History of Heredity* (Houghton Mifflin, 2001) and *Notes from the Shore* (Viking Penguin,

1995). She is writing a book about the human body (forthcoming from Houghton Mifflin, 2006). Ackerman is the recipient of fellowships and grants from the National Endowment for the Arts, the Radcliffe Institute for Advanced Studies, and the Alfred P. Sloan Foundation.

She writes: "The thrill of seeing whooping cranes in the wild during the research for this story was equaled only by meeting the man so instrumental in saving them, George Archibald. We met in the one-room log cabin of our mutual hero, naturalist, and writer Aldo Leopold. It was Leopold's *Sand County Almanac* that put conservation on the map of the American psyche; an essay in that book, 'The Marshland Elegy,' sparked Archibald's passion for cranes and my own desire to write. Listening to Archibald talk about whoopers and Siberian cranes in the cabin where Leopold wrote his little masterpiece was a peak experience."

PHILIP ALCABES is an infectious-disease epidemiologist, trained at Columbia and Johns Hopkins, with twenty-five years' experience researching hepatitis, AIDS, tuberculosis, and other epidemic diseases. He is associate professor of urban public health at Hunter College of the City University of New York, where he teaches courses on epidemiology, infectious diseases, and ethics. He has authored peer-reviewed articles featuring applications of statistical methods to elucidating the natural history of HIV infection, plus research on the social epidemiology of communicable diseases in urban settings. Alcabes's, work is on the history of contagion control and social construction of epidemics, a topic on which he is writing a book.

"Public anxiety about bioterrorism is less vivid in 2005 than it was a couple of years ago," he remarks. "'Smallpox' isn't a household word, or worry, now. But funding for 'biopreparedness' continues to rise, although how to defend against an imagined future scourge seems as unanswerable as ever. I find that public health policy is best when it pointedly addresses the diseases that beset the population here and now, the ones that scientists can study. My assessment of bioterrorism, public anxiety, and public health, first delivered in an open lecture at Hunter College, motivated this essay."

NATALIE ANGIER, whose science writing for the *New York Times* won her the 1991 Pulitzer Prize, started her career as a founding staff member of *Discover* magazine, where her beat was biology. In 1990 she joined the *Times*, where she has covered genetics, evolutionary biology, medicine, and other subjects. Her work has appeared in a number of major publications and anthologies, and she is the author of three books: *Natural Obsessions*, about the world of cancer research (recently reissued in a new paperback edition); *The Beauty of the Beastly*; and the national bestseller *Woman: An Intimate Geography*, published originally in 1999 and now available in paperback. She was the editor for Houghton Mifflin's *Best American Science and Nature Writing 2002*, and she is completing a book about what scientists wish everybody understood about science. She is also the recipient of the American Association for the Advancement of Science-Westinghouse Award for excellence in science journalism, the Lewis Thomas Award for distinguished writing in the life sciences, and the Freedom From Religion Foundation's "Emperor Has No Clothes Award," among other honors. Angier lives in Takoma Park, Maryland, with her husband, Rick Weiss, a science reporter for the *Washington Post*, and their daughter, Katherine Ida Weiss Angier.

"I first read about Jacqueline Barton and her research in a 1991 newsletter from Barnard College, our mutual alma mater in New York City," she remembers. "We were pictured side-by-side on the cover, I for having just won the Pulitzer Prize, she for having been named a MacArthur Fellow. My initial impulse was a little trillium of deadly sins: envy, a Pulitzer is swell, but it can't compete with what Roy Blount, Jr., so eloquently called 'those goddamned genius awards'; greed, her prize pot was, oh, two orders of magnitude bigger than mine; and vanity, hey, how come *she's* the one from the earlier graduating class, yet I'm the one with the nasolabial folds?!

"My secondary response was a bit more fruitful. I scanned the newsletter synopsis of Barton's work, on the biophysical structure of DNA, and found it alluringly abstruse. I figured that, being a female chemistry professor at a world-class science institute like Caltech, Barton surely was a member of the Zero to Ten Club, right up there with 'number of women heading Fortune five hundred companies,' 'number of women who have won an Oscar for Best Director,' and 'number of women who have so much as toyed with the idea of running for president of the United States.' I also figured that, as a graduate of my beloved tough-gal college, Barton must dislike life in the token booth enough to be a champion for one of my preferred causes, women in science—or at least to have a few tart thoughts on the theme. I decided to write a profile of her.

"Why it took me thirteen years or so to visit her, I can't say, but I can say I'm glad I finally did. Barton is one of the most expansive and least high-handed scientists I've ever interviewed. She genuinely believes that most people could understand most science if only they made the effort, and that what impedes them is not an innate lack of ability, or a defect in some 'math gene' or 'chemistry gene,' but the perverse pride so many members of the laity take in their ignorance of science, their willingness to shrug, 'It's all geek to me.' Barton shares with her fellow chemists a certain defensiveness about their field, and the way the public will automatically interpret the word 'chemical' to mean 'an unnatural substance' and chemists to be 'scientists who pollute the earth.' Everything in the world is made of chemicals! Barton cries. There's nothing as 'natural' as a chemical! Yet she has a sly wit and a buoyant pragmatism that keep her from tendentiousness or a tendency to self-combust. Over the years she's heard hundreds of people tell her the same thing: they 'flunked high school chemistry.' Everywhere, in every dale, an epidemic of Fs. Surely there must have been a few Bs and Cs to round out the grade curve, she says sardonically. Surely there must have been the occasional, shameful, unspeakable . . . A?

"Barton is indeed a rarity, on her campus or elsewhere. In the top fifty American universities, only 8 percent of the full professors in chemistry are women. At Caltech, which is as small as it is elite, there are a mere two hundred and seventy-five faculty members overall. Of those, only thirty-five, or less 15 percent, are female; and of the thirty-five, only Barton and sixteen others are full professors. True to my expectations, Barton has worked hard to help recruit more women into the ranks.

And true to that photograph of so many years ago, she remains implausibly unfurrowed. Must be that goddamn genius award."

ANDREA BARRETT is the author of five novels, most recently *The Voyage of the Narwhal,* and two collections of short fiction: *Ship Fever,* which received the National Book Award; and *Servants of the Map,* a finalist for the Pulitzer Prize. A MacArthur Fellow, she has also received fellowships from the National Endowment for the Arts, Guggenheim Foundation, and, in 2001–2002, was a Fellow at the Center for Scholars and Writers at the New York Public Library.

"'The Sea of Information' began as notes for a lecture I gave at Denison University in the spring of 2003," she explains. "For several years before that I'd been thinking about, gathering material for, and then beginning to write a novel set in a tuberculosis sanatorium in the northern Adirondacks during WWI. After a running start I'd been stopped dead in my tracks by events both external and internal; utterly flummoxed, for the first time in my life I was unable to write. Inevitably, I suppose, an examination of the reasons why became the subject of the talk, which David Baker and David Lynn, of *The Kenyon Review,* helped me develop more fully as an essay. As I tried to articulate some of the difficulties of conveying complex subject matter— science and medicine, in my case—in a fashion that speaks to the soul and the heart as well as to the mind, I relearned the lesson that all real writing comes from the same source. The essay, which I had started only reluctantly, turned out to be the path leading me back into the novel—now very different from what I first imagined. An experiment, like all the others."

DAVID BERLINSKI was born in New York City in 1942 and educated at the Bronx High School of Science, Columbia College, and Princeton University, from which he received his Ph.D. in 1967. He taught philosophy and logic at Stanford University during the 1960s, and during the 1970s worked as a management consultant with McKinsey and Company and as a senior quantitative analyst for the City of New

York. A professor of mathematics at the Université de Paris, Jussieu, in the late 1970s, Berlinski thereafter held research positions at the Institute for Applied Systems Analysis in Austria, and the Institut des Hautes Etudes Scientifiques in France. He has taught mathematics at a number of American universities. He is the author of *Black Mischief: Language, Life, Logic, Luck; A Tour of the Calculus; The Advent of the Algorithm; Newton's Gift; The Secrets of the Vaulted Sky;* and, most recently, *Infinite Ascent: A Short History of Mathematics.* Berlinski's books have been translated into ten languages. He lives in Paris.

"I wrote this essay as a part of a larger and far more ambitious project," he comments, "one devoted to a discussion of the origins of the mind, the origins of life, and the origins of matter. Although 'On the Origins of Mind,' deals with evolutionary psychology, my aim was never to heap abuse on a discipline that has already been more than heaped-upon. Quite the contrary. Evolutionary psychology has had a cultural impact—the stuff is quite literally everywhere—so considerable as to suggest that some effort should be made to judge its claims against the sober and demanding standards of the serious sciences. It is just when they *are* assessed in this way that those claims appear unfounded.

"And not only *those* claims. The human, and animal, mind represent things in the world that we can barely describe and that obviously we do not understand. No system of psychology has ever done anything more than suggest that there are mirrors behind the mirrors that we see."

WILLIAM J. BROAD is a senior writer at the *New York Times* and author or coauthor of six books, most recently *Germs: Biological Weapons and America's Secret War,* a number-one *New York Times* bestseller. For more than two decades, he has covered science for the *Times,* reporting on such topics as geology, astronomy, oceanography, biology, physics, astrophysics, space weapons, and nuclear arms. Mr. Broad has won two Pulitzer Prizes with *Times* colleagues, as well as an Emmy. He holds a master's degree in the history of science from the University of Wisconsin.

"I fell in love with earth science in 1993," he writes, "after cramming myself into a tiny submersible and dropping a mile and a half to the bottom of the Pacific to witness the birth of planetary crust. I wrote about the dive in a book, *The Universe Below,* and began doing articles for the paper on the earth's inner secrets. One told how scientists in 1995 succeeded in making the first computer simulation of field reversals. In June 2004, when a large effort at tracking such shifts got underway, I couldn't resist writing about the subject again."

K. C. COLE, according to her friend Dava Sobel, is "our ambassador to the realms of the exceedingly strange"—her favorite description of her work. Among her books are: *Mind Over Matter: Conversations with the Cosmos, The Hole in the Universe: How Scientists Peered Over the Edge of Emptiness and Found Everything,* and *The Universe and the Teacup: The Mathematics of Truth and Beauty.* She's written for *The New Yorker,* the *New York Times,* the *Washington Post, Esquire,* and *Newsweek.* She has taught at the University of California, Los Angeles; Yale; and Wesleyan. She has won the American Institute of Physics Science Writing Prize and the Los Angeles Times Awards for deadline reporting and explanatory journalism.

"Much of my writing focuses on the question, 'How do we know what we know?'" she says. "When we look out into the cosmos or into the quantum mechanical world of atoms, it's nearly impossible not to project our own experience into realms where they don't apply; we look into a mirror and think we're seeing the outside world. Astrobiologists who seek extraterrestrial life on other planets have it particularly hard, because their experience is limited to a single case: life on Earth. And so, they look for carbon and water—the stuff we're made of—in part because it may be the only kind of life they can recognize. But did life have a choice? Could it have been made of other ingredients? Such speculations inevitably circle back to an even more interesting question: Do we really know what life is? The answer is: probably not. Like so much of forefront science, the search for extraterrestrial life may turn out to be a quest for something we can't even conceive."

JENNIFER COUZIN was born in Montreal, Canada, and grew up in Toronto. While an undergraduate at Harvard, she steered clear of science classes before becoming fascinated by the history of science and selecting it as her major. Since 2002 she has been a staff writer at *Science* magazine, covering various issues in medicine and basic biology. Her work has also appeared in *U.S. News & World Report, Newsweek*, and the *Washington Post*, and other publications. In 2004 Couzin won the Evert Clark/Seth Payne Award, given annually to an outstanding science journalist age thirty and under.

"I lucked out with Lenny Guarente and David Sinclair," she reports, "because the two are such charismatic scientists and freely opened up to me. The scientific dispute between them—among the most complex biology I've ever covered—is still not resolved. But tensions have eased: Guarente and Sinclair now agree that both their theories might simultaneously be true. They sometimes eat dinner together with their families, and are coauthoring a scientific article. Sinclair told me recently that he's 'like any child who's left the nest—eventually, they come back.'

"While other biologists continue testing Guarente and Sinclair's disparate theories, the two have moved on to new pursuits: examining whether their aging theories in yeast hold up in mice, a key step that would link their work to human biology. Both, not surprisingly, are optimistic that they're on the right track."

MARK DOWIE teaches science reporting at the University of California Graduate School of Journalism. He is the former editor-at-large of *InterNation*, a transnational feature syndicate based in New York, and a former publisher and editor of *Mother Jones*. During his thirty years in journalism, Dowie has written over two hundred investigative reports for magazines, newspapers, and other publications. And he has written five books. His works have won seventeen journalism awards, including four National Magazine Awards (NMA) (he has also been an NMA finalist seven times), a George Polk Award, as well as citations from the National Press Club, Sigma Delta Chi, Project

Censored, the University of Kansas (William Allen White Gold Medal), and the University of Missouri (Penny-Missouri Award). In 1982 he was awarded the bronze medallion by Investigative Reporters and Editors (IRE), his fourth award from that organization. In 1992 Dowie received the Media Alliance's Meritorious Award for Lifetime Achievement, and in 1995 was awarded a Doctor of Humane Letters by John F. Kennedy University. He is now writing a history of conservation refugees.

"One story leads to another," he remarks. "I tripped over this one while writing about the patenting of genes and other life forms. Here was a biologist, Stuart Newman, who thought badly enough about the idea of creating human-primate chimeras that he filed for a patent to stop the invention of what he foresaw as an inevitable and horrific new life form. Since the story was published the U.S. Patent Office has denied Newman's patent. He plans to appeal the denial all the way to the Supreme Court, forcing the justices to deal with a question with which they are rarely faced: What is human?"

PETER GALISON is the Mallinckrodt Professor of the History of Science and of Physics at Harvard University. He was a John D. and Catherine T. MacArthur Foundation Fellow, and in 1999 won the Max Planck Prize. His books are *How Experiments End* (1987), *Image and Logic* (1997), and *Einstein's Clocks, Poincaré's Maps* (2003); coedited volumes, include *Architecture of Science, Picturing Science, Producing Art,* and *Scientific Authorship.* He coproduced a documentary film on the moral-political debate surrounding the H-bomb, *Ultimate Weapon* (2000), and is now working on *Secrecy,* about government classification.

He writes: "With a mythologized, older Einstein in mind, we think of this epochal physicist as disconnected from the practical world. But as a young man, Einstein was anything but that—he was an efficient and recognized patent officer, someone who testified as an expert witness and even took out his own patents. One area of his expertise was in gyrocompasses—devices that used gyroscopes instead of magnetic needles to find direction. Out of that work came a remarkable piece of science in which Einstein designed a new theory of magnetism, and even conducted extraordinarily delicate experiments. Einstein's practical and theoretical worlds were far closer than we might imagine."

As a medical and science writer for *Newsday*, New York City, LAURIE GARRETT became the only writer ever to have been awarded all three of the big "Ps" of journalism: the Peabody, the Polk (twice), and the Pulitzer. She is also the bestselling author of *The Coming Plague: Newly Emerging Diseases in a World Out of Balance* and *Betrayal of Trust: The Collapse of Global Public Health.* In March 2004 Garrett took the position of Senior Fellow for Global Health at the Council on Foreign Relations. She is an expert on global health with a particular focus on newly emerging and reemerging diseases; public health and their effects on foreign policy and national security.

ATUL GAWANDE is assistant professor of surgery at Harvard Medical School, assistant professor of health policy and management at Harvard School of Public Health, and a staff writer for *The New Yorker.* His book, *Complications: A Surgeon's Notes on an Imperfect Science,* was a finalist for the National Book Award. It has been published in sixteen languages and sold in more than seventy-five countries. Gawande and his family live in Newton, Massachusetts.

"Since writing this article on the daunting effort to eradicate polio," he reports, "whether it will succeed has not become any more certain. On the one hand, polio infections in India fell to fewer than on hundred and fifty cases in 2004. Yet in western Africa the epidemic exploded. There were seven hundred and eighty-nine identified in Nigeria alone, and the virus spread from there into a dozen countries where it had once been eliminated. We still face the possibility that the campaign will prove to be a benighted, enormously expensive failure—or our generation's greatest achievement."

JAMES GLEICK is an author, reporter, and essayist. His latest book, *Isaac Newton,* was a 2004 Pulitzer Prize finalist and a national bestseller, as were *Chaos: Making a New Science* (Viking Penguin, 1987) and *Genius: The Life and Science of Richard Feynman* (Pantheon, 1992). His other books, include *Faster: The Acceleration of Just About Everything* (Pantheon, 1999) and *What Just Happened: A Chronicle from the Electronic Frontier* (Pantheon, 2002). They have been widely translated abroad.

"Just as I was winding down from writing a biography of Isaac Newton," he explains, "my hometown library put on a glorious exhibition of Newtoniana—including manuscripts I had traveled across an ocean to see in darkened rooms and some I had never been able to see at all. The editors of *Slate* asked me to mosey over there and say what I thought. I was thrilled. But then I was disappointed too."

JEROME GROOPMAN holds the Dina and Raphael Recanati Chair of Medicine at the Harvard Medical School and is chief of experimental medicine at the Beth Israel Deaconess Medical Center. He serves on many scientific editorial boards and has published more than one hundred and fifty scientific articles. His research has focused on the basic mechanisms of cancer and AIDS and has led to the development of successful therapies. His basic laboratory research involves understanding how blood cells grow and communicate ("signal transduction"), and how viruses cause immune deficiency and cancer. Dr. Groopman also has established a large and innovative program in clinical research and clinical care at the Beth Israel Deaconess Medical Center. In 2000 he was elected to the Institute of Medicine of the National Academy of Sciences. He has authored several editorials on policy issues in *The New Republic,* the *Washington Post,* the *Wall Street Journal,* and the *New York Times.* His first popular book, *The Measure of Our Days* (Viking, 1997), explores the spiritual lives of patients with serious illness and the opportunities for fulfillment they sometimes find. In 1998 he became a staff writer in medicine and biology for *The New Yorker.* His second book, *Second Opinions* (Viking, 2000), addresses the complexity of navigating the uncertain world of medical diagnosis and treatment. His third book, *The Anatomy of Hope* (Random House, 2004), is a *New York Times* bestseller.

"Early one evening," he recalls, "I was walking back to my laboratory after seeing patients in the hospital. Some had cancer, others had AIDS. Some would improve and live; others would not. I asked myself 'What more can I do for them?' I pondered the tests I had ordered and the treatments I had initiated. There were no further clinical interventions that seemed urgent. Then, a thought appeared: I could try to give them hope. I was both exhilarated and disturbed by the idea. Hope seemed powerful and vital, yet fragile and possibly false. And so it can be.

"I decided to try to learn about hope, its place in medicine and biology. I have not stopped learning. For my patients and myself."

BEN HARDER writes for *Science News,* where he covers medicine and the environment. He began the writing life as a Southeast Asia correspondent for the *Let's Go* travel guides and later served as editor-in-chief of that series. He has also reported on education, travel, and health for *U.S. News & World Report* and on science for National-Geographic.com and other publications. In 2002 he visited East Africa on a Knight Journalism Fellowship at the Centers for Disease Control and Prevention. He studied biological anthropology at Harvard.

"Science is most intriguing, I find, when it initially seems counter-intuitive," he explains. "As medical tools, maggots attracted me precisely because they conjure up images of death and decay. 'Creepy Crawly Care' actually began life as part of another story, but I realized there was too much of what could be called 'gross medicine' to fit into just one piece. The rest of the material appeared in *Science News* in February 2002, under the title 'Germs That Do a Body Good: Bacteria Might Someday Keep the Doctor Away.'"

ROBIN MARANTZ HENIG has written eight books, most recently *Pandora's Baby: How the First Test Tube Babies Sparked the Reproductive Revolution.* Her articles have appeared in the *New York Times Magazine, Civilization, Discover, Scientific American,* and just about every woman's magazine in the grocery store. Since 1998 she has been on the board of directors of the National Association of Science Writers. Henig has received an Alicia Patterson Foundation fellowship and a

Sloan Foundation grant, and was a finalist for a National Book Critics Circle Award for *The Monk in the Garden: The Lost and Found Genius of Gregor Mendel.*

"I was worried about inadvertently taking a misstep in my article about the genetics of race," she says. "Race is such a loaded issue in this country, as is the concept of genetic determinism, and here I was about to tackle both. But my piece seems to have satisfied both sides in a fiery debate. The best part of writing it was conducting face-to-face interviews, more than two dozen, in which it was okay to acknowledge that my interviewees and I were different races, I'm white, and that we might experience the world differently as a result."

EDWARD HOAGLAND was born in New York in 1932 and sold his first novel, *Cat Man,* set in a circus, at the age of twenty-one. It is still in print. He began to write essays in his mid-thirties and has published many collections, spanning a love for both wilderness and metropolises, and attempting to plumb the chaos of current life.

"Human nature is part of Nature," he says. "I try to write about both. 'Small Silences' is of a piece with my memoir, *Compass Points,* published in 2001."

JIM HOLT, a former editor of *The New Leader,* writes about science and philosophy for *The New Yorker,* the *New York Times Magazine,* and *Slate.* He has taught mathematics at various universities and has been a journalist in residence at the Mathematical Science Research Institute at the University of California, Berkeley. He also contributes a weekly gossip column to the *New York* magazine.

He writes: "Listening to great physicists like Freeman Dyson, Stephen Weinberg, and Ed Witten think out loud about how the universe will end is about the most entertaining thing I have ever done as a journalist. The question elicits a wonderful combination of hard science, ingenious speculation, and existential irony. I am still wondering just how worried I should be about the 'dark energy' that seems to be slowly—very slowly—shivering the cosmos to pieces."

GINA KOLATA reports on science and medicine for the *New York Times*. The author or coauthor of six books; she has received numerous awards for writing about medicine, mathematics, and statistics; and, in 2000, she was a finalist for a Pulitzer Prize in investigative journalism. She studied molecular biology in graduate school and also has a master's degree in applied mathematics.

"I wrote this article for myself as much as anyone else," she says. "Why, I wondered, was stem cell research being portrayed as a choice between using adult stem cells or ones taken from embryos? Were the scientists themselves divided into adult versus embryonic stem cell camps by different views of the ethics of working with human embryos? Were research agendas being driven by ideology? But ideology was not the issue, at least at the labs I visited. Instead, it was the basic human instinct to let one good idea lead to the next and to explore surprising findings."

DENNIS OVERBYE is a science correspondent for the *New York Times* and the author of *Einstein in Love* (Penguin, 2001) and *Lonely Hearts of the Cosmos* (Little, Brown, 1999). He lives in Manhattan with his wife, the writer Nancy Wartik, and his daughter, Mira.

"If you ask an astronomer what's new in the universe these days," he remarks, "he or she will probably tell you about two discoveries, more or less at opposite ends of the scale. One is a fact of almost brute generality, namely that some antigravity, as bizarre as that sounds, seems to be speeding up the expansion of the universe; if this keeps up, the universe will be completely dark, empty, and cold in only a few trillion years. The other is almost cozy and cuddly by comparison, the discovery of more than one hundred planets circling other stars in our galaxy. The planets were all initially detected indirectly, but by the time this essay is published, astronomers will have announced that they have seen the light from some of them directly.

"This is huge news—even if the planets so far found are uninhabitable supergiants—if you want to find life in the universe, or to understand the context in which life started here, as most scientists believe it did—there are always a few fans for panspermia. By all accounts the heroic efforts of the last decade are only the beginning of the beginning of the quest for life-friendly abodes, if not life itself. NASA plans to spend hundreds of millions on it, if the agency's science programs survive the shift to Moon-Mars exploration.

"The idea that the universe could suddenly seem so friendly on a relatively small scale and so ominous and forbidding on the largest, grandest scale is intriguing, and even suggestive. If you ask a cosmologist what keeps him or her awake at night, one of the answers will be the question of why the antigravity, known perhaps too familiarly as 'dark energy,' has come to dominate the universe at about the same time as life evolved. This strange question is likely to be one of the vital and controversial centers of intellectual debate for the foreseeable future."

DAVID QUAMMEN travels on assignment for various magazines, most often to jungles, deserts, and swamps. His accustomed beat is the world of field biology, ecology, evolutionary biology, and conservation, though he also occasionally writes about travel, history, and outdoor sports. His work has appeared in *Harper'* magazine, *National Geographic,* the *Atlantic Monthly, National Geographic Adventure, Outside,* the *New York Times Book Review,* and other journals. His books, include *The Song of the Dodo; Monster of God;* and a spy novel, *The Soul of Viktor Tronko.* He lives in Montana with his wife, Betsy Gaines, a conservationist; their large furry dog; and a modest supply of cats.

He writes: "This essay had its origin not in my head but in the head of Bill Allen, editor-in-chief of *National Geographic* magazine at the time, who one day decided that first, before he retired, he would run a piece on the evidence for evolution; and second that 'Was Darwin Wrong?' would be its title. A third decision, somewhere along the way, was that he wanted me to write it. The call came at an opportune time for me, when I was just beginning a year's work on a larger Darwin project—a short book, cast as a biographical essay, for the Great Discoveries series published by Atlas Books and W. W. Norton. The piece did run, in the magazine, under Bill's title, which served effectively as a hook to bring in readers from both sides of the question. But for this anthology, I've given it a different title, slightly less teasing. In early 2005 'Was Darwin Wrong?' won the National Magazine Award in the Essay category.

"The magazine assignment was especially welcome because it offered me an excuse to visit several eminent evolutionary biologists—notably Douglas Futuyma, Niles Eldredge, and Ian Tattersall—and engage them on the subject. Neither Eldredge nor Tattersall are mentioned in the piece, but I had a wonderfully genial two-bottle-of-wine lunch with them at a bistro in New York, preceded and followed by hours of serious talk in their offices at the American Museum of Natural History. Taped to a shelf above Eldredge's desk, I noticed an old snapshot of him as a grad student, along with his pal and eventual coauthor of the theory of punctuated equilibria, the late Stephen Jay Gould, both of them looking young, thin, and eager. Another snapshot showed Eldredge with his collection of five dozen cornets, each showing a different stage in the evolution of that musical instrument. Affixed to the glass door of one of his bookshelves was a bumper sticker: HONK IF YOU LOVE DARWIN. Concrete human details and experiences like these—the lunch, the snapshots, the personal office decor—are part of the reason I don't care to rely much on the telephone-interview method of journalistic research."

OLIVER SACKS is the author of nine books, including two collections of case histories, *The Man Who Mistook His Wife for a Hat* and *An Anthropologist on Mars*, in which he describes patients struggling to live with conditions ranging from Tourette's syndrome to autism, parkinsonism, phantom limb syndrome, schizophrenia, and Alzheimer's disease. He has investigated the world of deaf people in *Seeing Voices*, and a rare community of colorblind people in *The Island of the Colorblind*. His most recent books are *Oaxaca Journal* and the autobiographical *Uncle Tungsten: Memories of a Chemical Boyhood*.

"As a boy, I was fascinated by the Periodic Table," he recalls, "and was thrilled to learn, only a few months after World War II ended, that four new elements—elements 93, 94, 95, and 96—had been discovered, the first ones beyond uranium. Sixty years later, in February of 2004, when the isotopes of elements 113 and 115 were created, I again got very excited, and immediately started thinking about their possible properties. So when I got a phone call from David Shipley at the *New York Times* asking if I would like to write something about these new elements, I said I would love to. Most of my writing is accomplished with huge labor, multiple drafts, and false starts, but 'Greetings from the Island of Stability' seemed to leap out spontaneously, perhaps because, for me, it echoed the elemental joys of so long ago."

MARK SOLMS was born in 1961. He moved from Johannesburg to London in 1989, completed his doctoral research in neuropsychology in 1991, and qualified as a psychoanalyst in 1994. He published *The Neuropsychology of Dreams* in 1997 and founded the interdisciplinary International Neuro-Psychoanalysis Society, along with its journal, *Neuro-Psychoanalysis,* in 1999. He won the American Psychiatric Association's International Psychiatrist award in 2001. He returned to his native South Africa in 2002, where he holds the chair of neuropsychology at the University of Cape Town.

"The invitation to write this article came as a welcome surprise" he says. "It has not always been easy to convince my professional and scientific peers that psychoanalysis, for all its faults, is a valuable and important tool. Over the past few years, there have been growing indications that my view might finally be prevailing. I interpreted the invitation from *Scientific American* in this light. For me, the central problem of current neuroscience today is to understand how and why the brain is not the same as any other bodily organ. The brain alone has subjective capacities; it can tell us how it feels. Psychoanalysis still offers science the best available methodological and theoretical framework for exploring and understanding the unique perspective on nature that this capacity provides."

ELLEN ULLMAN is the author of *The Bug: A Novel* and *Close to the Machine,* a memoir about her twenty years of experience as a software engineer. Her essays about the social and emotional effects of computing have appeared in *Harper's* magazine, *Salon, The American Scholar, Wired,* and the *New York Times.*

She writes: "When I first starting working as a computer programmer, in the late 1970s, only the geekiest of my colleagues referred to their memories as 'data bank' or to their activities as 'multitasking.' Over the decades, I watched in alarm as these technical terms became common, even prescriptive, until it seemed that most people in the developed world saw their deepest human faculties in terms of computers. I think I began writing about technology to do battle with that vision of human beings as information processors: to remind people that computers were created as our complements—good at things we're bad at—not as our models. I turned to the subject of food—the most fleshly of human needs; like sex, reeking of desire—because I thought, Here is a part of human existence that can't possibly be programmed. I found I was wrong."

FRANK WILCZEK is a theoretical physicist aspiring to become a natural philosopher. He likes all kinds of puzzles, reads voraciously, and plays at music and swimming. He is open to further suggestions and adventures. Wilczek has won prizes both for science and for writing, including recently the Nobel Prize. He is a product of the New York City public schools.

Regarding his essay he has contributed the following poem:

To grasp for the unknown, I must extend my range.
To live amidst the tiny, to bring the distant near
I must adjust my focus, to keep my vision clear.
Turning my eyes homeward, I saw a truth most queer:
As the strange becomes familiar, the familiar seems quite strange.

Permissions

A Note from the Series Editor

Submissions for next year's volume can be sent to:

Jesse Cohen
c/o Editor
The Best American Science Writing 2006
HarperCollins Publishers
10 E. 53rd Street
New York, NY 10022

Please include a brief cover letter; manuscripts will not be returned. Submissions made electronically are also welcomed and can be e-mailed to jesseicohen@netscape.net.

SCIENCE WRITING AT ITS BEST

THE BEST AMERICAN SCIENCE WRITING 2004
Dava Sobel, Editor
Jesse Cohen, Series Editor
ISBN 0-06-072640-7 (paperback)

THE BEST AMERICAN SCIENCE WRITING 2003
Oliver Sacks, Editor
Jesse Cohen, Series Editor
ISBN 0-06-093651-7 (paperback)

THE BEST AMERICAN SCIENCE WRITING 2002
Matt Ridley, Editor
Jesse Cohen, Series Editor
ISBN 0-06-093650-9 (paperback)